Library and Archives Canada Cataloguing in Publication

Mobile Nation: Creating Methodologies for Mobile Platforms
Edited by Martha Ladly and Philip Beesley

This anthology is based on the Mobile Nation conference (2007) but is distinct from
the conference proceedings published under the same title.
Includes bibliographical references and index.
ISBN 978-0-9780978-4-4

1. Design and technology.
2. Wireless communication systems.
3. Mobile communication systems.
4. Design, Industrial.

I. Ladly, Martha, 1955-
II. Beesley, Philip, 1956-

TS171.M63 2008   621.382   C2008-901298-4

CDRN/RCRD

Canadian Design Research Network
Réseau canadien de recherche en design

mdcn

ONTARIO COLLEGE OF ART & DESIGN

Networks of Centres of Excellence
Reseaux de centres d'excellence

 Social Sciences and Humanities
Research Council of Canada

Conseil de recherches en
sciences humaines du Canada

 Canadian   Patrimoine
Heritage   canadien

ARCHITECTURE
WATERLOO ARCHITECTURE CAMBRIDGE

INTERACTIVE
PROJECT LAB
LABORATOIRE POUR
PROJETS INTERACTIFS

MOBILE NATION

Edited by
Martha Ladly
Philip Beesley

Riverside Architectural Press

The Mobile Digital Commons Network and Mobile Nation

Sara Diamond
Ontario College of Art & Design

Michael Longford
Concordia University

Mobile Nation partner the Mobile Digital Commons Network (MDCN) connects research, arts, and industry focused on mobile, wireless, digital technologies in Canada. Funded by Canadian Heritage through its New Media Research Networks Fund, the goal of the network over two rounds of funding has been to facilitate interdisciplinary research and innovative industry development; foster cultural production and public participation in culture through the use of mobile technologies; strengthen relationships to wilderness, heritage, and urban parks through enriched experiences in these spaces; and develop forward-thinking Canadian policy on wireless technologies. It has invented applications and technologies that facilitate the design of mobile content and experiences.

The MDCN is a collaborative research project originally launched by Concordia University and the Banff Centre's Banff New Media Institute, and now including York University and the Ontario College of Art & Design. The projects bring together an interdisciplinary group of computer scientists, engineers, technology and communication scholars, artists, and designers, as well as commercial partners in Canada and abroad, in order to explore wireless communication in the context of urban studies, prototype development, commercial applications, art installations, participatory public authoring, design methodology, and social research.

The MDCN has created policy reports on Canadian mobile infrastructure and investment. It has developed a comprehensive guide to the use of participatory design and user-based evaluation in mobile research and design. The network has undertaken two conferences, and published extensively on mobile research through *Wi*, its digital journal, and in other academic venues.

Mobile Nation celebrated the completion of a three-year research project for MDCN. It offered diverse experiences from keynotes to workshops, panels to student symposia, parties to exhibitions. We hope you enjoy the celebration, in retrospect, through this publication.

Design Research and the Mobile Experience

Douglas MacLeod and Robert Woodbury
Canadian Design Research Network

The Canadian Design Research Network (CDRN) is pleased to offer its support for Mobile Nation. The interactive technologies embedded in the concept of mobility suggest a new generation of responsive designs that will transform the world we live in. In this context, the CDRN provides a forum for sharing information, personnel, and resources connected to design research in this area and acts as a vehicle for disseminating that research through events such as Mobile Nation.

Funded through the government of Canada's Networks of Centres of Excellence New Initiative program, the CDRN is a pan-Canadian consortium of academic institutions, government agencies, and industrial partners that is promoting good design as the best means to improve the social, physical, and economic well-being of all Canadians.

In particular, we are committed to supporting and connecting the activities of graduate students in all disciplines who are exploring the design issues associated with these mobile technologies. GPS, Radio Frequency Identification, and Wi-Fi networks all define a new kind of space that demands the kind of innovative design research presented by Mobile Nation.

Through conferences such as Mobile Nation, the CDRN aims to foster the development and maturation of the discipline of design by networking across both distance and discipline. To this end it includes faculties of design, architecture, engineering, computer science, environmental design, construction, and landscape architecture. We work together to provide the research that will transform the practice of design in the twenty-first century.

Mobile and interactive technologies are clearly part of that transformation and in concert with other CDRN research themes such as digital fabrication, collaborative visualization, and advanced design technologies we hope to address critical issues such as productivity and sustainability, and to lead in the creation of economic advantages and environmentally responsible solutions in Canada.

www.cdrn.ca.

Contents

Mobile Nation
creating methodologies for mobile platforms

Sara Diamond and Martha Ladly
Ontario College of Art & Design

Dialogues about commercial applications for the mobile platform, new technological innovations, and market development often occur in separate forums unrelated to scholarly dialogues about the sociology of mobile use and creative practices outside of the contemporary commercial market. Researchers, developers, and investors in these areas all have specific approaches that they seldom have an opportunity to share. For this reason, this Mobile Nation anthology brings together very different communities to share a dialogue about their methodologies and approaches to this growing field. Our hope with it is to foster an increased capacity to work together to build the creative, technical, and social capacity that mobile computing will bring. The Mobile Nation conference, upon which this anthology is based, provided an innovative and rigorous context within which to consider appropriate research methods, including their epistemology and their application. It also offered an opportunity to develop new and appropriate methodologies for this challenging, complex, and important field of inquiry.

Researchers have created platforms such as cellular telephones, MP3 devices, and PDAs, and these have become leading consumer products. They have invented communications technologies such as Wi-Fi and Bluetooth, sensor systems such as Radio Frequency Identification (RFID) and GPS, and networks (such as personal area networks). These in turn enable the development and combination of new tools to create content. Examples include new forms of fashion that can respond to social context, environment, or wearer, and new architectural expressions such as interactive billboards. Other researchers are analyzing this growing mobile phenomenon from a social and business perspective.

The wider context of increasingly social media sets the stage for the current growing collaboration of artists, scientists, designers, and engineers. The challenges of designing mobile experiences and technologies are many, requiring teams with a range of disciplinary knowledge and skill, as well as the ability to manage constant changes in platforms, complexity of programming languages and challenges of building content appropriate to mobile devices. There are added challenges when the content is location-based rather than generic, or collaborative rather than individual. This framework requires the expansion of hardware and the development of interactive capacity, hence significant code development and the creation of interactive content, all within a context of understanding and anticipating users' cultures and practices.

The Mobile Nation conference created an environment for researchers, companies working in mobile content and technology development, and users of the resulting products to share their interests and their actual approaches to conducting research in this fast-changing field. For example, effective research for a mobile experience that occurs outdoors requires that a mixed team of content creators, such as game designers, documentary producers, or creators of mobile walking tours, work closely with engineers to brainstorm, build, and test their designs with users on site. They must consider the actual physical location as their set, with all of its possibilities and technical limits, instead of working in the studio. They must understand the limits of GSM coverage and GPS accuracy and undertake actual engineering at the location of the intended experience.

The Mobile Nation anthology surveys five broad themes:
1. Participatory culture, ethnography, participatory design, and the end-user
2. Creating for the multi-platform context and challenges of media and place
3. Pervasive and social computing
4. Mobile communication and education
5. Engineering methodologies and solutions meeting humanities and social science approaches

More specifically, Mobile Nation considers how we can meet the challenge of integrating variable media (television and mobile for example) into the specifics of place; how we design for technical platforms where we expect to communicate and share content, with ethnography and participatory design providing specific tools for researchers and industry alike; how mobile media serve the changing context of informal and formal education; how ubiquitous computing (or ubicomp) results in the reorganization of social experiences; and how, in turn, this field can benefit from mobile capacities, that is, how new designs engage physical environments, whether built or worn. Finally, it asks how social scientists' methods of analysis of behaviours, particularly of adoption and usage patterns, can be combined with engineering solutions to build better research and innovation, and more marketable products.

Working Methods

By providing a focus on 'methodologies,' the conference threw an international light on mobile research and commercialization, afforded opportunities for high-level exchange between national and international players, and highlighted the work of leaders in the field of mobile scholarship. For this reason Mobile Nation attracted a diverse group of individuals from different professions, such as architects, educators, broadcasters, designers, representatives from infrastructure companies who provide the networks to deliver content, health researchers, advertisers, and leaders from technology companies. In our commitment to innovation across these fields, Mobile Nation offered a workshop on sensors and interactive technology and an introduction to a new prototyping tool developed by engineers and designers at the Mobile Digital Commons Network (MDCN), the Mobile Experience Engine (MEE).

The field of mobile design requires new levels of both research collaboration and engagement with end-users. Mobile experience design demands the full integration of participatory design into the research and innovation chain. Participatory design emerged in northern Europe during the sixties as a means to engage workers in planning and implementing technological change within their workforce.[1] It emerged side-by-side with participatory action research, where communities of use or study are integrated throughout the research process.[2] Since then, participatory design has caught on as a means to engage end-users in the process of creating both technologies and experiences. Participatory design strategies suggest that researchers have the responsibility to engage in the research project with their subjects in order to transform the matter that they are acting on.[3] Participatory design is also described as 'user-centred design,' recognizing the importance of the end-user and incorporating them into the design process from the conception to the evaluation of the final product, although 'the forms and degree of involvement vary (representative or direct involvement, consultants, or collaborators)... [It] aims at involving future users of a computer-based system in decisions during system development.'[4]

Participatory design processes are deeper than market research methods that concentrate on surveys or one-time focus groups, although these methods can provide a tool for usability testing. In participatory design, researchers engage participants in context, finding metaphors for the experience or technology that they are building in order to engage participants' imaginations or understand relevant social phenomena. A core group may work with researchers throughout the research project to achieve a depth of engagement. Participants in the design process may feel that they should dictate the framework of the final product. Participatory design walks a fine line, as designers and inventors still need to mobilize their professional knowledge to create new experiences.[5] Participatory design can be used to effectively balance the views of designers, engineers, and end-users.

Participatory design advocates have devised a variety of techniques to facilitate the communication and testing of new technology possibilities to users. These techniques include the use of mockups and role-playing activities, as well as technologically aided methods such as the use of photos, images, videos, or animations to stimulate the patterns of interaction with a new interface or system.[6] Participatory design methods include ethnographic study of users in their working environment or during field

[1] **T. Winograd,** *Bringing Design to Software* (Boston: Addison-Wesley, 1996); **A.G. Bjerknes and T. Bratteteig,** 'User Participation and Democracy: A Discussion of Scandinavian Research on Systems Development,' *Scandinavian Journal of Information Systems,* 7 no. 1 (1995) : 73–98.

[2] **P. Reason and H. Bradbury,** eds., *Handbook of Action Research: Participative Inquiry and Practice* (New York: Sage, 2002).

[3] **J. Greenbaum and M. Kyng,** *Design at Work: Cooperative Design of Computer Systems* (Hillsdale: Erlbaum, 1991).

[4] **Bjerknes and Bratteteig,** 'User Participation and Democracy'

[5] **R. Wakkary, K. Newby M. Hatala, D. Evernden, and M. Droumeva,** 'Interactive Audio Content: The Use of Audio for a Dynamic Museum Experience Through Augmented Audio Reality and Adaptive Information Retrieval,' in *Museums and the Web 2004: Selected Papers,* 55–60 (Toronto: Archives and Museum Informatics, 2004); **D. Schuler and A. Namioka,** *Participatory Design: Principles and Practices* (Hillsdale: Erlbaum, 1993).

trials that simulate the experience that a tech-nology is designed to address.[7] Recent trends in art and design research encourage workshop activities and the making of artifacts such as collages, mind maps, and models. Structuring and presentation of the resulting data is a key part of the researcher's work.

A core tool of participatory design is brainstorming. Brainstorming is a value-neutral, conceptual free-for-all where discussion members are encouraged to put their wildest ideas on the table. Once a topic or problem to solve is chosen, all ideas are encouraged without criticism, in order to allay fear and loose the imagination. So-called 'blue-sky' notions can turn out to be appropriate solutions.

Another, more recent technique, bodystorming, draws from performance art. In bodystorming, participants use physical improvisation to explore forms of interaction, emotional content, and relationships between individuals and groups. Place-storming or location-storming is a technique first formalized by Urban Tapestries, a London-based organization that designs participant-driven, location-based historical experiences in specific neighbourhoods.[8] They design in situ with their participants in order to bound the imagination of users and designers within an actual location.

The design charette is an extended approach to these processes that designers and architects use. The charette combines the language-based approaches of brainstorming with the expectation that an actual series of designs and even prototypes will emerge. The charette format allows for the collaborative participation of artists, designers, engineers, and stakeholders.[9]

Iterative design is an engineering method aligned with extreme software programming in which an engineering solution is built in small increments, tested, and improved by a team through the process of development. This method allows constant adjustment of the technology to the actual circumstances of application. It is less likely to result in a technology that has no relationship to the needs of users or the context of use.[10] It is a technical proximate to participatory design in an engineering context. When problems are found in user testing, as they will be, they must be fixed. This means design must be iterative: there must be a cycle of designing, testing and measuring, and redesigning repeated as often as necessary. Participatory design, bodystorming and location-storming, charettes, and iterative design are a range of methods that were explored at Mobile Nation.

Research

The Mobile Nation conference celebrated the completion of a three-year research project, the MDCN, funded by Canadian Heritage through its New Media Research Networks Fund. The Ontario College of Art & Design (OCAD), as a member institution of the MDCN, is a leader in the development of mobile technologies and content. The MDCN con-

6 **M. Muller and S. Kuhn,** eds., participatory design special issue, *Communications of the Association for Computing Machinery*, 36 no. 4 (1993).

7 **C. Wasson,** 'Collaborative Work: Integrating the Roles of Ethnographers and Designers,' *Human Organization* 59 no. 4 (2002): 377–388.

8 **Proboscis,** 'Bodystorming,' http://research. urbantapestries.net/ bodystorming.html.

9 **M. Aurand,** 'What is a Charette?' www.library.cmu.edu/ Research/ArchArch/Charette/ what.html.

10 **S.N. Wakeford and E. Churchill,** 'Framing Mobile Collaborations and Mobile Technologies,' in *Wireless World: Social and Interactional Aspects of Wireless Technology*, eds. B. Brown, N. Green, and R. Harper. (London: Springer, 2001).

nects research, arts, and industry focused on mobile, wireless, and digital technologies in Canada. The network facilitates research and innovative industry development; fosters cultural production and public participation; and develops forward-thinking policy on wireless technologies. This project has seen the creation of numerous exciting prototypes for new forms of content and experiences that occur in urban and national parks, using mobile devices as a key component. The Mobile Nation conference was an opportunity for participants to explore experiences that included annotated and illustrated walking tours, historical ghost stories, private-to-public ephemeral graffiti, and collaborative sound games.

This conference provided an opportunity to disseminate the MDCN's collective efforts in the creation of the MEE, a technology that will greatly assist in future design of mobile games for cellphones. Developers can benefit from the knowledge that our research teams gained. MEE has been created to radically simplify the process of creating and managing media-rich, interactive mobile applications, and in particular location-based applications using GPS and peer-to-peer applications such as Bluetooth. By removing most of the engineering from the design cycle and enabling designers to create complex applications using simple XML language, MEE makes rapid prototyping a reality for mobile applications and takes mobile application product development into new domains. At the Mobile Nation conference, the MEE was introduced to the community, and developers and designers had the opportunity to create their own mobile applications.

By applying humanities and social science knowledge and methods to the analysis of the emerging mobile content and platform worlds, we can innovatively engage design theory, communications studies, social geography, cultural studies, and ethnographic research methodologies. In bringing experts and scholars together, Mobile Nation enables more comprehensive, effective, and integrated research and inquiry.[11]

The Canadian Design Research Network (CDRN), a National Centre of Excellence for the dissemination of and training in design research, was a partner in the creation of the Mobile Nation conference and publications. The CDRN presented an interactive sensor and technology workshop, a special poster and symposia events for students, industry panels, and outreach to communities that can make use of the discoveries at the confe-rence through Mobile Nation publications.

The conference subtext, 'creating methodologies for mobile platforms,' intentionally uses the term 'creating' as a double entendre. Mobile research methods must allow collaborative 'creating' on the part of designers, engineers, and users. At the same time, mobile research requires the 'creating' of unique, cross-disciplinary methodologies and tools that will enable all manner of new innovation in the field. We hope that you find this collection of essays connected to the Mobile Nation conference to be useful and stimulating.

[11] **K. Cohen,** 'Translation: Sociology: Design' (paper presented as a visiting lecturer to the Oxford Internet Institute, Oxford, UK, 2003); **K. Cohen,** 'Applying Collaboration Theory to Social Spaces' (presented at BRIDGES Conference II, Banff Centre for the Arts, Banff, AB, October 4–6, 2002).

Mobile Nation
key themes and key thinkers

Martha Ladly
Ontario College of Art & Design

This anthology reflects five key themes that present central challenges and opportunities for researchers, designers, and creators working in the mobile realm today. Many of the thinkers who presented papers and work at the Mobile Nation conference generously expanded their conference presentations for inclusion in this book. These works from Mobile Nation's thinkers are gathered together here because their insights allow readers to bridge design practice with research creation and theory. The authors are eminent scholars, creators, designers, and practitioners who share an impressive range of experience and expertise, an infectious curiosity, and an ability to think across disciplinary, cultural, and technological boundaries. I believe that our writers have provided direction on difficult questions in ways that are clear and understandable. At the same time, they pose new and sometimes unconventional questions, and they suggest new directions for creative, technical, and business practice and scholarship.

Mobile Nation's, expert leaders were invited to initiate dialogue on central topics in mobile research and readers will find foundational essays by these leaders opening the discussions within. Their essays lay the ground for a depth of descriptive, technological, project-based reports and theoretical papers that expand upon the following key themes:

1. Participatory culture, design, and ethnography
2. Creating for the multi-platform context
3. Pervasive and social computing
4. Mobile communication and education
5. Engineering meets humanities and social science

The range of discussions in Mobile Nation suggests both the complexity of mobile media and the need to approach research and innovation in the field from many different perspectives. I hope readers will discover that the insights offered in Mobile Nation excite and enable us to move more deeply into the mobile context.

I would like to acknowledge the important contributions of my co-editor Philip Beesley, the careful editorial work of Leah Sandals, the coordination of this project by Siobhan O'Flynn and Eric Bury, and the work of all of the authors.

Participatory Culture, Design, and Ethnography

All mobile technology researchers and industry professionals are also end-users. Mobile Nation provided a large audience with a lively and engaging opportunity for professional, student, and non-professional end-users to compare user contexts and to share ideas and information. With a focus on ethnographic research methodologies, Mobile Nation's expert speakers and scholars made strong links to achieve our objective of moving forward the study and application of ethnographic methods within the fields of art, design, and communication research. These links between fields include participatory and user-centered design, charettes, ethnographic research, participant observation, iterative design, and improvisation. Presentations in this realm gave researchers and industry professionals the opportunity to share their ideas and innovations, to 'try out' ideas, and to receive valuable feedback from end-users, who became participants in the design process.

Questions of particular importance exist: How does the participatory design model work within the mobile context? What can we learn from user adoption, adaptation, and change? How do social-geographic and ethnographic research approaches contribute to our knowledge? And what approaches can be taken to increase accessibility of mobile content for a variety of users? The social, co-operative use of mobile telephones breaks from earlier, individual-centred design protocols, and the potential new uses of the technology demonstrate the ways in which technologies can be 'reimagined' and 'repurposed' by new user communities.

Mobile Culture and Perceptions of Future Mobile Applications
cultural values and usage patterns

James E. Katz
Rutgers University

The first handheld cellular telephone call was made in Manhattan in April 1983; in less than a mere twenty-five years, about 2.8 billion (out of 6.6 billion) people worldwide have become mobile phone subscribers. The mobile phone has enjoyed a more rapid adoption rate than any other *(fig.1)* technology, the vaunted and highly successful Internet being no exception. In some countries, there are more cellphone subscribers than there are people. For instance, Italy has 124 subscribers per hundred people; the Czech Republic, 115; Hong Kong, 123; and Israel, 112. Not only is the spread of the technology amazing, but so too is its intensity of usage. For instance, in the US the mean daily use per subscriber is about two hours.[1] In 2005, over a trillion messages were sent using short messaging service (SMS or text messaging), a medium that had not existed commercially a decade earlier. That roughly translates into about 1.3 SMS messages per subscriber per day.

The way people spend their time on a daily and structural basis has of course been modified by this huge upsurge in mobile communication. For instance, in the US in June 2006, subscribers consumed about 850 billion 'minutes of use.'[2] That is to say, the 150 million plus subscribers in that month used more than two hours per day of mobile phone minutes. On the one hand, it is safe to say that many users consume far fewer minutes. On the other hand, one must also consider that, for all those subscribers who use, say, fifteen minutes per day, someone else had to use a lot more minutes to get the mean number to balance out at around 120 minutes. Perhaps it is even useful to think in total numbers: more than 1.5 million 'person-years' were consumed talking on the mobile phone, an unimaginably vast amount of human endeavour.

Public space and behaviour have also been altered by the arrival of mobile media. As but one piece of evidence, a 2007 study at Rutgers University of 'tele-density' found that about 25 percent of people on campus were using mobile devices at any one time. These levels have been on an upward trend *(fig.2)*. Based on tens of thousands of observations, another point worth observing is that while in 2005 more males than females were users, the trend has reversed.[3] While obviously not a representative sample of the world, these findings add further confirmation that there is widespread and heavy use of these technologies.

facing page
1 **Mobile penetration rate**
per hundred inhabitants

2 **Mobile communication**
usage on a college campus
by gender

Mobile Penetration Rate per 100 Inhabitants

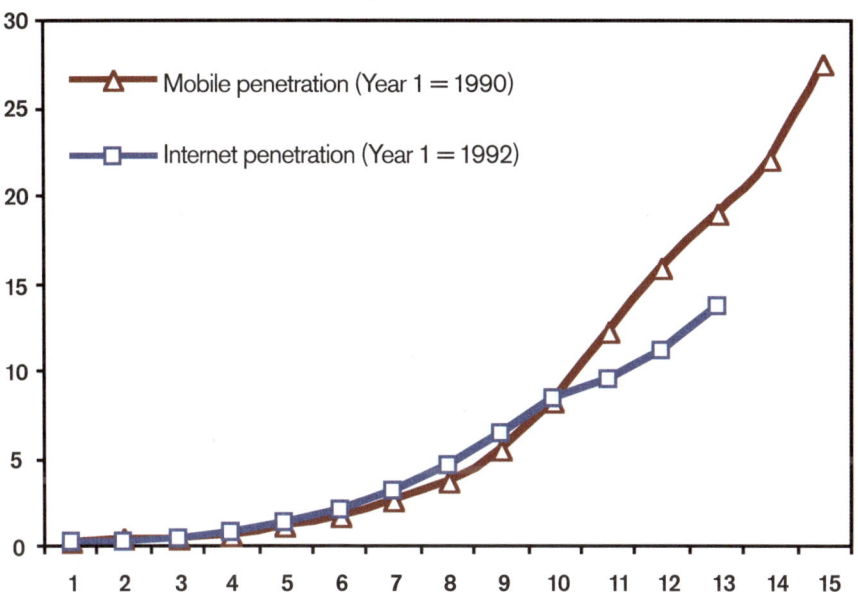

Mobile Communication Usage on a College Campus by Gender

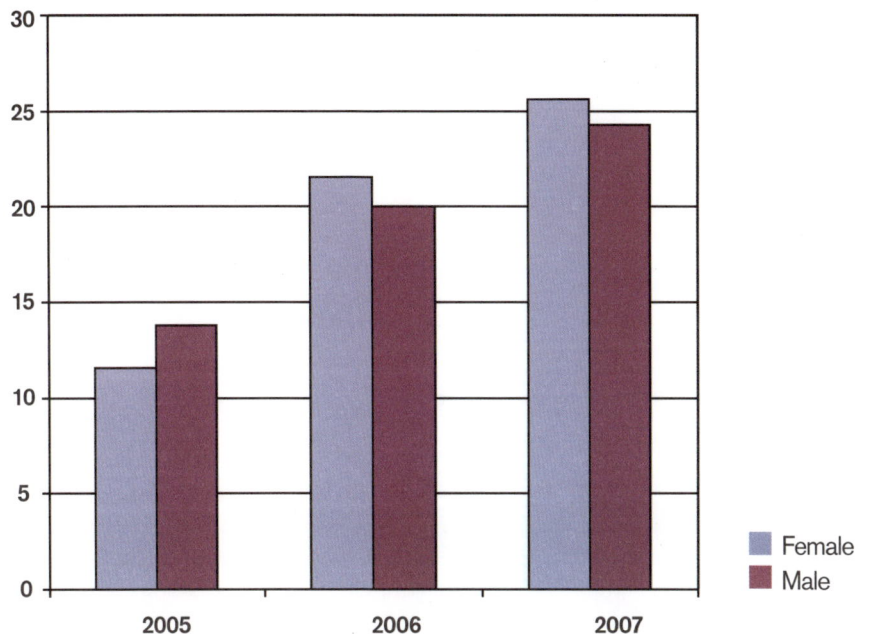

A host of ancillary services have become available. For example, in the US about one out of five cellphone subscribers downloads ring tones. Their average cost is about US$2.32 each.[4] The North American ring tone download market yields about half a billion dollars a year.[5] Geo-positioning, location monitoring, games, and real-time television broadcasts are but a few of the current services, and two-way video link services are proliferating.

I introduce these figures and points to suggest just how invested many people are in their mobile phones. However, people are also amazingly creative in imagining and creating alternative uses and meanings for their devices. For instance, they use them for religious services, and for reminding themselves to conduct their religious obligations, as well as for receiving religious messages, fortune telling, and showing off. Of course, it is obvious that they also use them for more quotidian purposes of interpersonal communication, seeking help in emergencies, and conducting business. There are manifold illustrations of the ways these devices have been put to use. They range from offering prayers at temples to looking for dates. Mobile phone accoutrements include hand-carved mahogany cellphone holders for when the device is not in use, and special silk bags for carrying the mobile device around. Phones themselves may be jewel-encrusted or bear personalizing stickers and other ornamentations.

Certainly there are important moral implications to the widespread and innovative use and abuse of mobile phones. Sometimes there are the equivalent of 'moral panics' over the use of a device. Only this year, Iran's telecommunications ministry announced that it would begin filtering immoral Multimedia Messaging Service (MMS) messages. According to a May 2007 report by Reuters,[6] the Supreme Council of the Cultural Revolution, which was established after Iran's 1979 Islamic revolution, has ordered the telecommunications ministry to purchase the equipment needed to prevent any misuse of MMS. This filtering of MMS is being done 'in order to prevent possible misuse of MMS, immoral actions, and social problems.'[6]

At the same time, my reading of press reports suggests that despite the fact that Internet cafes in the Gaza Strip and Afghanistan are frequent targets of attack by Islamic fundamentalists, the mobile phone towers have not been struck. That may be because technology is seen as vital to all sides. Moreover, it may be that attacking the mobile phone infrastructure would alienate the affected people. These topics cannot be explored here, but are given as illustrations that may suffice to highlight the highly symbolic and operational investment people make in the cultural reproduction of mobile telecommunications technology.

In order to understand better what interests people might have in new services involving mobile phones, my team at the Rutgers Center for Mobile Communication Studies and I conducted a public opinion telephone survey of people in the US aged eighteen years and older. We did this in late February and early March 2007; over the course of several weeks we gathered the opinions of a random sample of 1,404 people. We asked a series of questions

facing page

3 **An outsized mobile phone** in South Chicago serves as an eye-catching advertising device. Reminiscent of the memorable eyeglasses billboard in F. Scott Fitzgerald's novel *The Great Gatsby* the mockup also invites passersby to consider the mobile phone's looming surveillance potential or, at the very least, its omnipresence.

4 **Mobile communication** greatly improves lives for pedicab drivers in New York City, as it does elsewhere around the world. Although an ultramodern technology, the mobile phone fits with virtually every lifestyle including some of the most labour-intensive ones.

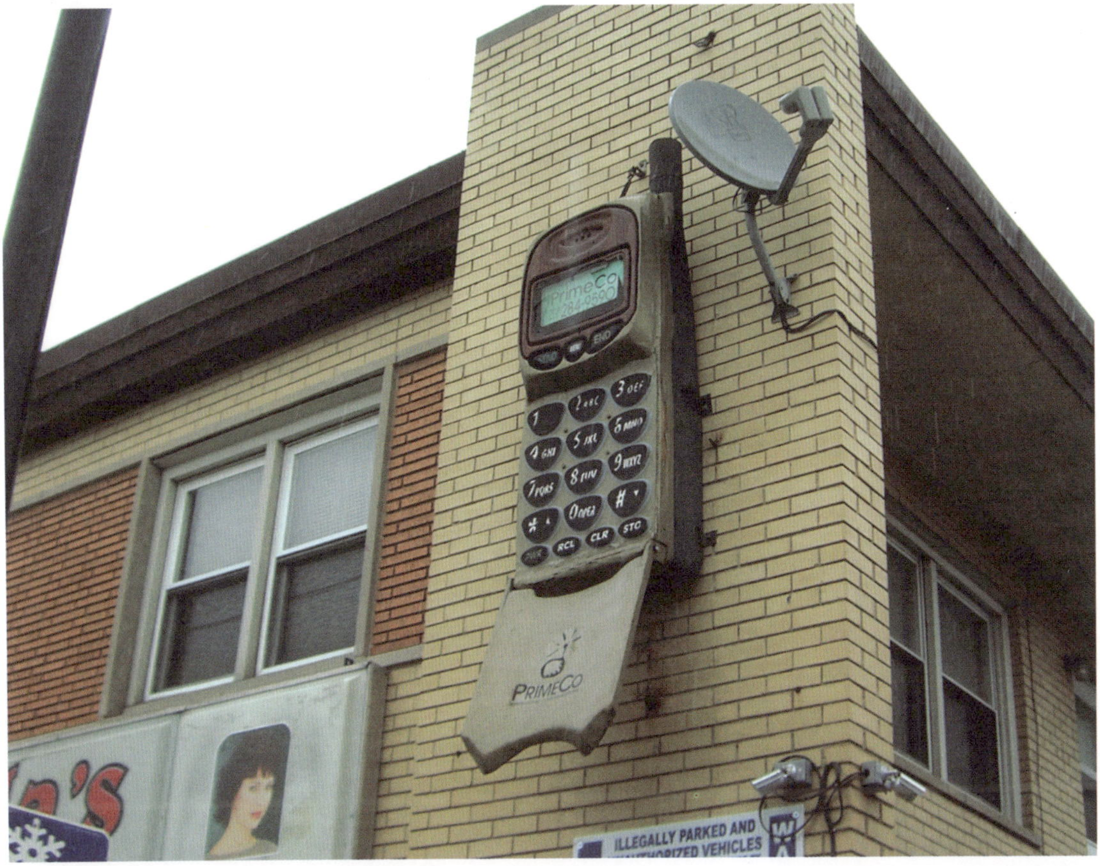
PrimeCo
PRIMECO
ILLEGALLY PARKED AND

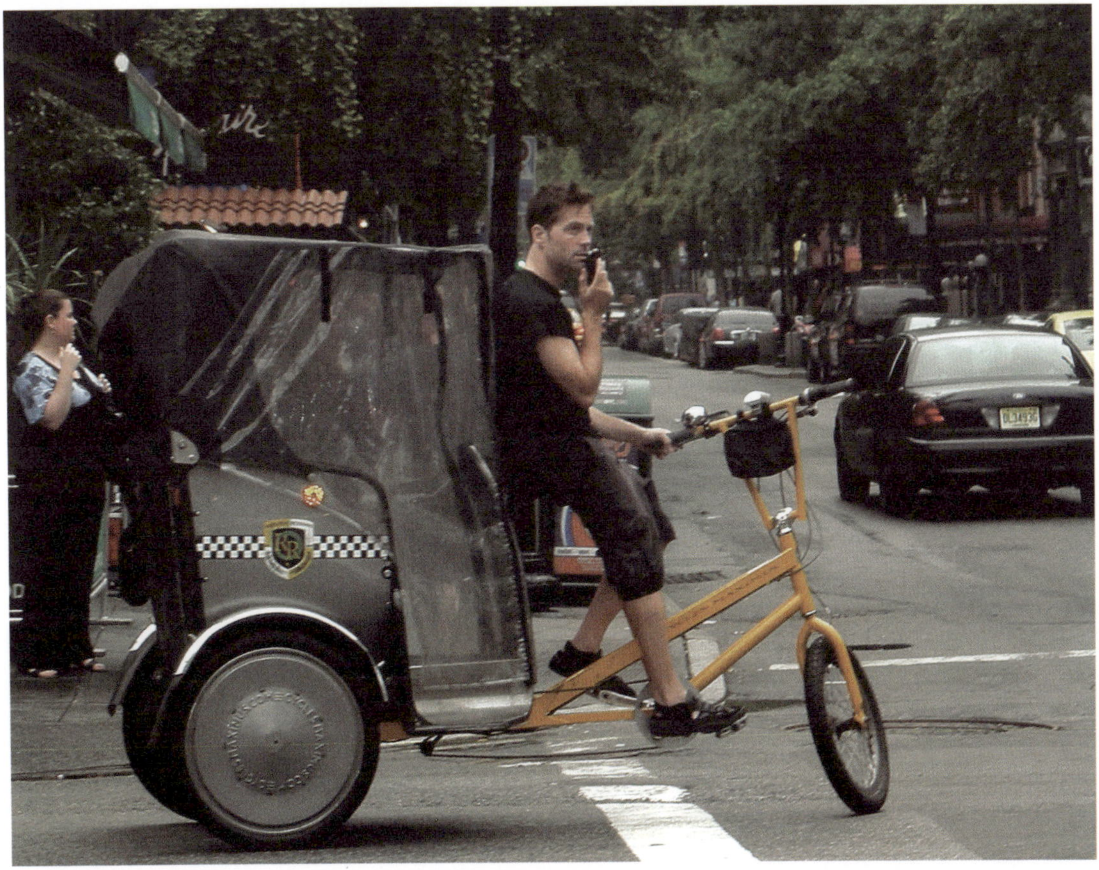

about mobile phone usage, including whether users, using a five-point Likert scale, thought the following services that could be offered over a mobile phone were a good idea or a bad idea:

- getting ads about products or services you might be interested in
- getting notices about bargains offered by local merchants in areas where you happen to travel
- watching TV
- playing games
- getting brief notices sent to your phone about important health information
- having a hotline to a doctor at any time
- finding directions when you're lost

5 **The mobile phone** can both provoke and allay ennui and makes it easy for users to be alone together, as is shown in this image captured in a restaurant a few blocks from Chicago's Art Institute. The scene reminded me of Edward Hopper's iconic *Nighthawks* which is displayed at the Art Institute.

What we found was that there was very little intense interest in any of the services except having a hotline to a doctor or getting directions if the user was lost. These latter two services can be readily appreciated as essential to people's well-being, so having a positive outlook towards such services is certainly understandable. Also not particularly surprising is that buying products or getting ads from local merchants were not viewed very favourably. But what was surprising to us was that other potential services which are generally viewed as fun also failed to receive high interest ratings. These were activities such as watching television, getting discount coupons, or playing games.

We also wanted to see whether privacy played an important role in interest in services. To do this, we asked people first if they were interested in being able to locate their friends using the mobile phone. While there was some interest in this, we followed that question with its reciprocal: would they be interested in a service that allowed friends to keep track of them? In this case, interest was lower. We can hypothesize that the reduced interest was a result of the relationship: people like to know where others are, but are less comfortable with others knowing how to locate them.

A few words are in order concerning these results. First, these are just snapshots in time. Just because someone is not interested in a service at present does not mean that a person will not be deeply interested in that service in the future. Second, the services were not explained to the respondent, so not much thought was likely given to the answer. Rather, they are 'off the cuff' responses and do not necessarily reflect a deeper understanding of what might be involved in such a service. Nonetheless, overall, there is less interest in the services than one might have originally expected.

So a few conclusions are in order. First, people are incredibly engaged with their mobile communication devices and services. While this is patently clear to any casual observer, it is useful to see some concrete measures of the phenomenon. We have provided a figurative hand-wave at some global and local data concerning the great success many nations have had at becoming 'mobile nations.' Second, many people discover elaborate and interesting new ways to use their personal technologies. We can sketch some of the innovative ways people use and also embellish their cellphones, which can include their use as transcendental tokens or symbols of lifestyle. Third, by taking a snapshot at one point in time, we can see that many new services are not likely to be an easy sell to a large proportion of the public. This, of course, does not mean that over time the public will not want the services. Rather, add-on services are likely not to be following the same rapid trajectory that was experienced by voice services, and may even have a more difficult time succeeding in the North American market than text messaging. Finally, this research suggests that small-scale experimental manipulation of wording can reveal deeper structural aspects of communication asymmetries. Or, in other words, by structuring questions in certain ways, as we have done in the example of who knows what about where one is, relative to the mobile phone, we can see that interpersonal privacy is not considered equivalent, but rather has asymmetric qualities. One apparently wishes to reduce one's visibility to one's friends even while wanting greater visibility of one's friends to oneself. People seem to prefer knowing where others are without revealing where they themselves might be. Thus, prompted by the questions we asked, people provide insights about relationships between mobile communication and the structure of people's lives—perhaps without intending to reveal them.

Notes

[1] CTIA, 'Wireless quick facts: December 2006,' www.ctia.org/advocacy/research/index.cfm/AID/10323

[2] CTIA, 'Wireless quick facts.'

[3] Yi-Fan Chen and Katie M. Lever, 'Teledensity: A Study of Gender Differences in the Use of Mobile Communication Technology on a College Campus.' (presentation, International Communication Association, San Francisco, CA, May 24, 2007).

[4] Siklos, Richard. 'Media frenzy: It's Like Selling Meals by the Bite. And it May Work,' New York *Times*, November 13, 2005. www.nytimes.com/2005/11/13/business/yourmoney/13frenzy.html?ex=1289 538000&en=0e1aa4f8560a975e&ei=5088&partner=rssnyt&emc=rss.

[5] 'Nielsen to Measure the Mobile Media Consumer; More than 33 Million Persons Used Mobile Web and 8 Million Persons Viewed Mobile Video in the Past 30 Days,' *PR Newswire US*, June 6, 2007, archived at www.nielsen.com/media/pr_070606.html; 'BMI Backs Off Ringtone Revenue Estimates, Citing Mature Market,' *Telecomweb*, April 13, 2007.

[6] 'Iran to filter "immoral" mobile messages,' Reuters.com, April 28, 2007, www.reuters.com/article/technologyNews/idUSDAH83913820070428.

[7] The Likert scale runs from strongly agree through to agree, neutral, disagree, strongly disagree, or similar uni-dimensional attitudinal valences.

Towards Issues-based Art and Design Research

Anne Galloway
Carleton University

What follows is a brief, and necessarily incomplete, first attempt to move some of my sociological research on mobile publics into the realm of art and design research. While questions of public participation, public space, and public technologies are well-known discourses in the development of wireless, mobile, and context-aware technologies, little systematic attention has been given to what actually constitutes the publics at hand. Issues-based art and design research can be seen to involve bringing people and things together around shared concerns—and I would like to suggest that it is one means by which we can 'spark a public into being.'[1]

Building on early twentieth-century debates between Walter Lippmann and John Dewey on 'the problem with the public,' contemporary cultural theorists Noortje Marres and Bruno Latour have recently argued for a reassessment of what we mean by 'public' and the politics of everyday life.[2] Rather than assuming a pre-existing public, or sphere awaiting (re)activation, Latour and Marres describe emergent and temporary assemblages of people interested and implicated in particular issues. Comprising a multitude of different individuals, these kinds of publics rally force by coming together around shared objects and concerns in contingent and inconsistent ways. Sociologist Mimi Sheller further argues that 'it is the capacity for coupling and decoupling in various ways that enables social action and the emergence of persons... Mobile publics can perhaps best be envisioned as capacitators for moving in and out of different social gels, including the capacity to take on an identity that is able to speak and to participate in specific contexts.'[3]

The notion that a differentiated public arises around shared objects of interest also appears in Elias Canetti's descriptions of feast crowds, and this sense of 'public' involves coming together in ways that are much more fluid and flexible than those associated with a rational public sphere: 'Many prohibitions and distinctions are waived... There is no common identical goal which people have to try and attain together. The feast is the goal and they are there... People move to and fro, not in one direction only. The things which are piled up, and of which everyone partakes, are a very important part of the density; they are its core. They were gathered together first, and only when they were all there did people gather round

facing page
1 **Pieter Bruegel**
fragment from *The Fight Between Carnival and Lent*, 1559, oil on oak panel
118 x 164.5 cm
Kunsthistorisches Museum
Wien, Vienna

PARTICIPATORY CULTURE, DESIGN, AND ETHNOGRAPHY

them.'[4] In his call for a new politics of things, Latour also argues that convergence may be more important than consensus: '[E]ach object gathers around itself a different assembly of relevant parties. Each object triggers new occasions to passionately differ and dispute. Each object may also offer new ways of achieving closure without having to agree on much else. In other words, objects—taken as so many issues—bind all of us in ways that map out a public space profoundly different from what is usually recognized under the label of "the political".'[5]

So what can art and design researchers make of all these notions of mobile publics? Starting with current understandings of mobile publics and public mobilities, including the manifestly technological, one option is to evaluate the capacity of people and things to move in and out of different contexts and identities. In order to do so, I believe that practitioners need to articulate and develop methods that are able to flex as much as the uncertainty, inconsistency, and instability of our particular situations and concerns demand. But perhaps most importantly, art and design researchers might first turn to particular issues or objects around which people may gather.

In order to move towards this kind of issues-based research practice, several existing approaches to, and methods for, collaborative work appear promising. Adapting Bill Gaver and colleagues' cultural probes to be used 'internally' rather than 'externally' could be one way to effectively bring a diverse project team together. [6] Another is to engage each other in what Jane McGonigal calls immersive and collective play.[7] Creative ways of generating ideas and documenting activities within and across fields can also be found in Proboscis' bodystorming, *StoryCubes*, endless landscapes, and everyday archaeology techniques.[8] All of these methods emphasize combinations of embodied experience, material and symbolic culture as means by which we form different identities and manoeuvre different situations. Although each method has traditionally been used as a means of working with users, participants, and other publics, they also provide means by which we can produce our own reflexive and local knowledges in the process.

But I also believe that we need to become more open to risky situations and relations, and to imagine different kinds of public politics, ethical research, and design practices. Issues-based art and design approaches will require flexible and temporary spaces of convergence, and practitioners may need to adjust the scale of their work. But ultimately, the greatest challenge may very well be learning when to intervene technologically, when not to, and re-evaluating what we mean by successful 'public' research in these domains.

Notes

1 Noortje Marres, 'Issues Spark a Public into Being: A Key but Often Forgotten Point of the Lippmann-Dewey Debate,' in *Making Things Public: Atmospheres of Democracy*, ed. B. Latour and P. Weibel (Cambridge: MIT Press, 2005), 208–217.

2 See Marres, 'Issues Spark a Public' and Noortje Marres, 'Public (Im)potence,' *Open* 11 (2007): 78–81. See also Bruno Latour, 'From Realpolitik to Dingpolitik, or How to Make Things Public,' in *Making Things Public: Atmospheres of Democracy*, ed. B. Latour and P. Weibel (Cambridge: MIT Press, 2005), 14–41.

3 Mimi Sheller, 'Mobile Publics: Beyond the Network Perspective,' *Environment and Planning D: Society and Space* 22 (2004): 49–50.

4 Elias Canetti, *Crowds and Power* (New York: Farrar, Straus & Giroux, 1998), 62.

5 See Latour, 'From Realpolitik to Dingpolitik,' 15.

6 Bill Gaver, T. Dunne, and E. Pacenti, 'Design: Cultural Probes,' *Interactions* 6 no. 1 (1999): 21–29.

7 Jane McGonigal, 'This Is Not a Game: Immersive Aesthetics & Collective Play' (paper presented at the Digital Arts & Culture Conference, Melbourne, Australia, May 9–23, 2003).

8 Proboscis, 'Bodystorming experiences,' http://proboscis.org.uk/bodystorming/, 'StoryCubes,' http://proboscis.org.uk/storycubes/, 'Endless Landscapes,' http://proboscis.org.uk/endlesslandscapes/, and 'Social Tapestries: Jenny Hammond School,' http://socialtapestries.net/jennyhammond/ydayArchaeologyReport06.pdf.

Morality, New Technology, and Engagement

Suzanne Stein
SMARTlab

As extensions of ourselves, technologies often open new ways of perceiving, being, and acting in the world. But what sort of world do we want? And what do we want our technologies to do for us? Foresighters are no longer simply developing responses to changing social, technological, and economic markets, but are using their methods to actively shape a desired future as cultural modellers. But with the democratization of the means of production and rich, global mobile and locative communications, posing these questions seems incumbent on all of us—we all have a stake in the continuous production and recreation of the world we live in. This paper proposes that producers begin a lively and dynamic exchange on the issue of professional ethics as codified moral stances.

There is a growing concern and public discourse on information communication technologies (ICTs) and their impact. This is international in nature and was characterized by the two successive UN summits on the information society in 2003 and 2005. With the global reach and widening digital divide, it has been necessary to pause, reflect, and try to change the course of what might seem like an inevitably closed-door, corporately monopolized space of communication. This reflection has been embodied by UN task force the World Summit Awards (WSA) that opened an international competition on ICT meaningfulness, awarding best practice in content and creativity to a selection of works nominated by close to two hundred countries worldwide.

There were forty awards given at each summit—all of these on display and circulating in an international road show designed to provoke deeper discourse and ignite the global imagination about what our ICTs might be.[1]

The awards were not oriented explicitly towards features and functions—nor did they necessarily prioritize the issue of real impact. Metrics of adoption, use, and mobilization can be dealt with elsewhere. The central issue was one of aesthetics: producer creativity and the contents of these works. Aesthetics are useful in providing us with the foundation of relevant discourse, delivering us principles for judgment and creation, giving us insight into what might be meaningful innovation.

The stakes are high. ICTs do not simply open up new avenues of communications, they create new ways of perceiving, new ways being in our world, and new situated knowledges. At the most profound level, they usher in issues of ontology and epistemology, which as a society we must explore and try to apprehend the consequences of.

facing page
1 *Flutter Fugue* performance by SMARTlab, UEL. SMARTlab brings artists, technologists, and performers together within a social inclusivity agenda, exploring the body, interactivity, and expression.

This philosophical and social role of new art and entertainment forms has been the subject of inquiry in media studies with each successive wave of expressive form. Marshall McLuhan was famous for his assertion that the 'medium is the message' and he wrote of the change in sense perception that our new technologies fostered.[2] And before this, with the rise of mass communications (particularly film), the Frankfurt School and associated academics, such as Walter Benjamin, debated their social impact—implicitly concerned with the role of fascism and media. Walter Benjamin wrote about the inherent change in 'apperception' and hoped these new media would put society 'under a microscope,' provoking our critical faculties.[3] Herein lies the role of the artist: to provoke reflection and provide the subject, space, and moment of contemplation.

Present-day concerns still reflect the subject of inquiry between media, being, and changing social relations that our new technologies afford. Of note is Roger Silverstone's call for reflection on the rise of global, mass, and personal communication channels, and the positioning of morality as a key concern. Here he posits morality as an orientation and procedure towards a notion of the 'good' and ethics as the codification of those practices. Not unlike Benjamin, he is implicitly casting the possibility of media practice as a counterpoint, or potential opposing force, to present-day international political havoc.[4] He too calls for reflection on our media creation and consumption patterns, hoping for the mobilization of a greater sense of morality and a change for the better in the creation and recreation of the world we inhabit.

Not surprisingly, the practice of understanding change and making sense of the future is becoming commonplace; foresighting as a discipline is back in vogue. It has a tendency to appear in times of uncertainty, change, and fierce competition. And more than playing a passive stance of determining a course of action in light of change, foresighters are increasingly trying to shape (the force and direction of) change itself. And this time, the artist and the ethnographer are appearing within this practice as gratifying bedfellows in this endeavour.[5]

And perhaps what is more striking than the rise of formal professional practice oriented towards shaping a better (or preferred) future is the popularization of such practice; the cultural modeller is a role we all play. Afforded by our new technologies, making an imprint on the world is increasingly quotidian. The spirit of playful engagement with the world is a new characteristic afforded by, for example, new technologies embedded in our architecture, wearables, and installations. The past few years have also marked the beginning of the end of the division between physical and digital realities. These are intricately, interstitially intertwined and manifest themselves in our activities of, for in object hyperlinking and the potentially totalizing effect of the 'Internet of things,' the notion of an 'overlay' of the digital seem archaic. Further, it will raise the stakes on the impact of the role ICTs in our lives, altering again our parameters of 'being' and 'knowing.' The idea that we individually adopt or use technology is itself passé; we participate and create with and from them.

facing page
2 [murmur] by Gabe Sawhney and Shawn Micallef, initially conceived at the Canadian Film Centre Media Lab, transforms city streets for *flâneurs* with mobiles accessing personal local stories

PARTICIPATORY CULTURE, DESIGN, AND ETHNOGRAPHY

3 **World Summit Awards Gala**
2005, Tunis

More than ever, our new technologies allow for a personal, direct, and immediate impact on the world. And be one an artist, producer, or person in the street, it seems that our aspirations for technology can be potentially manifest in the construction and participation in expressive new technology forms. As modellers we need to move conscientiously. However, the producer or artist role is still the primary engineer towards creating the frames, spaces, possibilities, and avenues of potential. Perhaps it is time to codify moral stances into (professional, personal, and collective) ethics. Similar to the impetus and rationale for the UN's WSA, accommodating aesthetics (as emerging new principles of creation and judgment) which are particular to our new social-technical spaces are in need of being forefronted in dialogue and debate. The kinds of narratives, the way in which our technologies situate us, the diversity of engagement that they may offer, and the issue of inclusion or exclusion are the stuff of morality—of orientation and procedure with which the world is cast. The global imagination has been prodded open, and hopefully through our discursive avenues and with the spirit of play and exploration, we will guide ourselves towards diverse, meaningful, and contemplative spaces of action.[6]

Notes

[1] See mission and the announcement of finalists at the World Summit Award website, www.wsis-award.org.

[2] M. McLuhan, *Understanding Media* (London: Routledge, 1964), 1–21.

[3] W. Benjamin, 'The Work of Art in the Age of Mechanical Reproduction,' in *Illuminations*, ed. H. Arendt, (New York; Schocken Books, 1969), 211–244. Also see T. Adorno and M. Horkheimer, 'The Culture Industry: Enlightenment as Mass Deception,' *Dialectic of Enlightenment*, (New York: Seabury, 1972), 120–167.

[4] R. Silverstone, *Media and Morality* (London: Polity, 2007), 7. '[I]t is because the media provide, with greater or lesser degrees of consistency, the frameworks (or frameworlds) for the appearance of the other that they, de facto, define the moral space within which the other appears to us, and at the same time invite (claim, constrain) an equivalent moral response from us, the audience, as a potential or actual citizen.'

[5] Some of this new collaboration is fostered by large institutions with the necessary resources and vision. Philips Design's Josephine Green, director of trends and strategies, has been one who has helped to spearhead these collaborations. Members of Nokia Corporation's Insight and Foresight Group and areas of Nokia Design, which I was a part of, were also investigating the possibilities of these collaborations. Large business aside, these collaborations are becoming commonplace in a variety of institutions both academic and commercial.

[6] Our labs, institutions, and conferences are a good starting point for collective inquiry, stimulation, and reflection. During my Mobile Nation talk, I presented case studies of the Canadian Film Centre Media Lab and University of East London's SMARTlab as embodying moral agendas through their institutional mandates. Labs such as these create a range of works, which are sometimes refreshingly contentious and defended by the explicit ethics of those producers. Debate on ethics and morality came to the fore at the Mobile Nation conference, possibly as an outcome of the successful range of work and thought presented.

facing page
4 **SEED** by the SEED Collective and CFC Media Lab, where a collectivity of people can grow a forest through their mobiles. This program works to create physical trees in areas in need of reforestation as a result of the individual, digital efforts.

5 **SEED Collective** in partnership with Seventh Generation at Expo East, Baltimore, Maryland, October 2006

PARTICIPATORY CULTURE, DESIGN, AND ETHNOGRAPHY

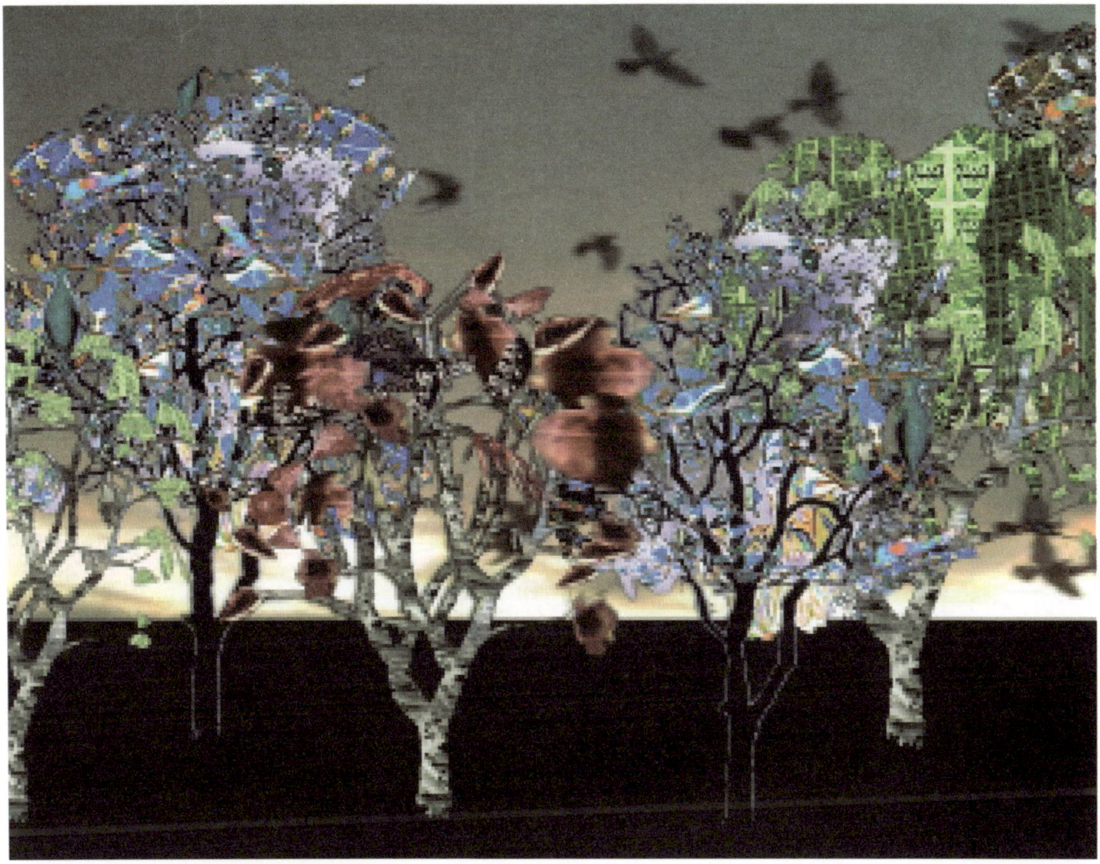

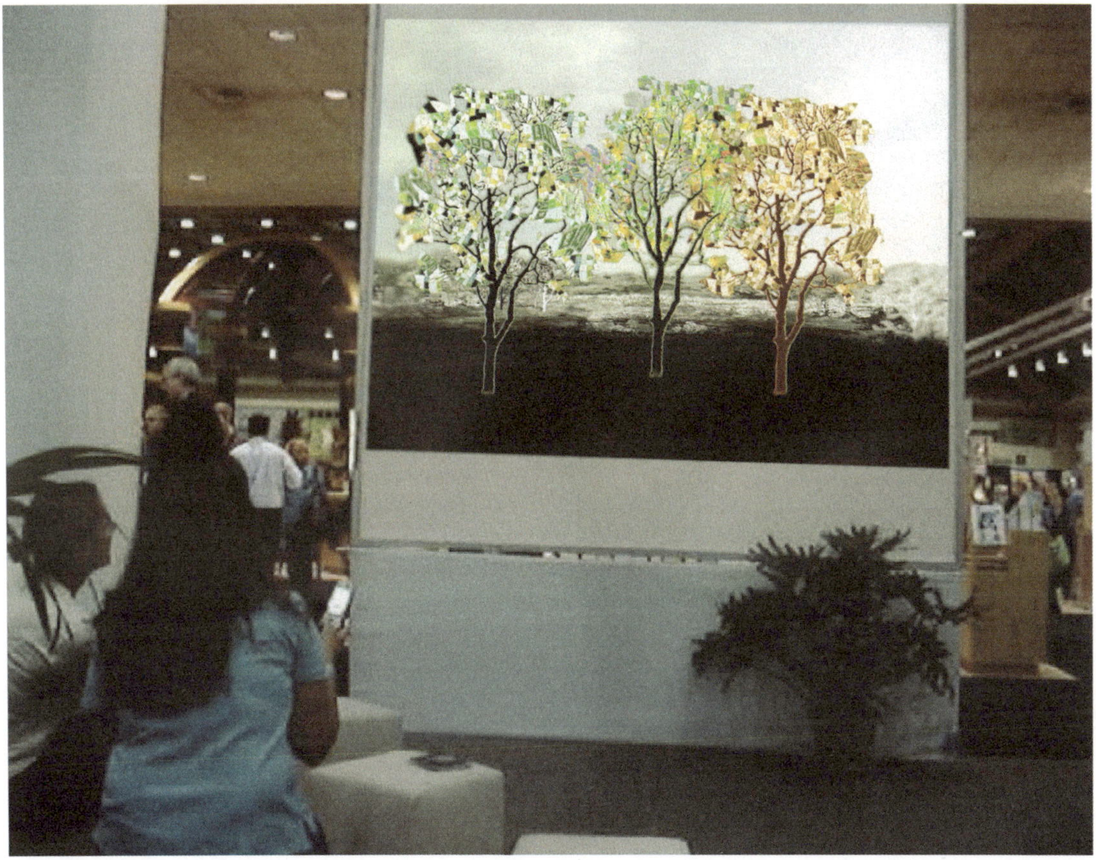

Play as Research
the iterative design process

Eric Zimmerman
Gamelab

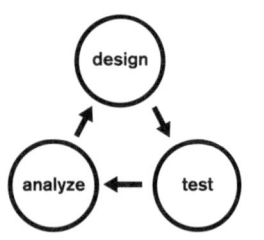

1 **The iterative design process**

Needs and Pleasures

Design is a way to ask questions. Design research, when it occurs through the practice of design itself, is a way to ask larger questions beyond the limited scope of a particular design problem. When design research is integrated into the design process, new and unexpected questions emerge directly from the act of design. This essay outlines one such research design methodology—the iterative design process—using three game projects with which I have been involved (SiSSYFiGHT 2000, Loop, and LEGO Junkbot).

The creation of games is particularly well-suited to providing a model of research through design. While all forms of design stem from both practical and sensual aims, game design is particularly skewed towards the creation of delightful experience, rather then the fulfillment of utilitarian needs. Although it is true that we can create and play games for a particular function (to exercise, to meet people, to learn about a topic), by and large, games are played for the intrinsic pleasures they provide.

As a form of designed 'delight,' the process of interacting with a game is not a means to an outside end, but an end in and of itself. It is this curious quality of games that makes them wonderful case studies for design research through the process of design. As a game evolves (through the iterative process outlined below), it defines and redefines its own form and the experiences it can provide for players. Through the iterative play of design itself, entirely new questions can come into being.

Iteration Iteration

Iterative design is a design methodology based on a cyclic process of prototyping, testing, analyzing, and refining a work in progress. In iterative design, interaction with the designed system is used as a form of research for informing and evolving a project, as successive versions, or *iterations*, of a design are implemented.

Test; analyze; refine. And repeat. Because the experience of a viewer/user/player/etc. cannot ever be completely predicted, in an iterative process design decisions are based on the experience of the prototype in progress.

facing page
2-3 **SiSSYFiGHT**

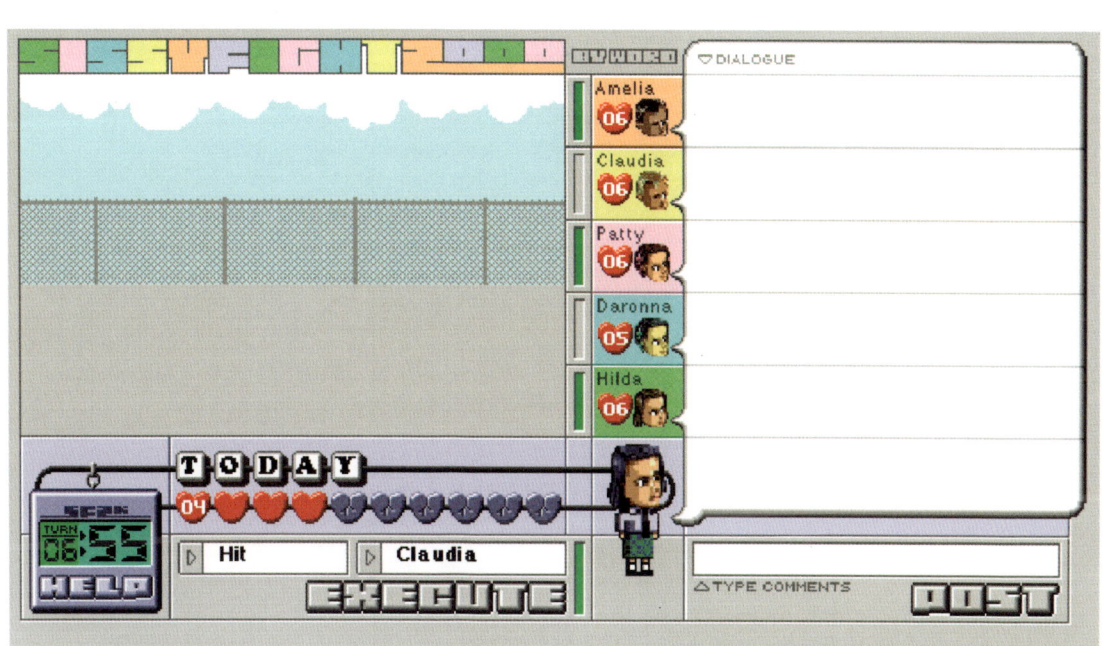

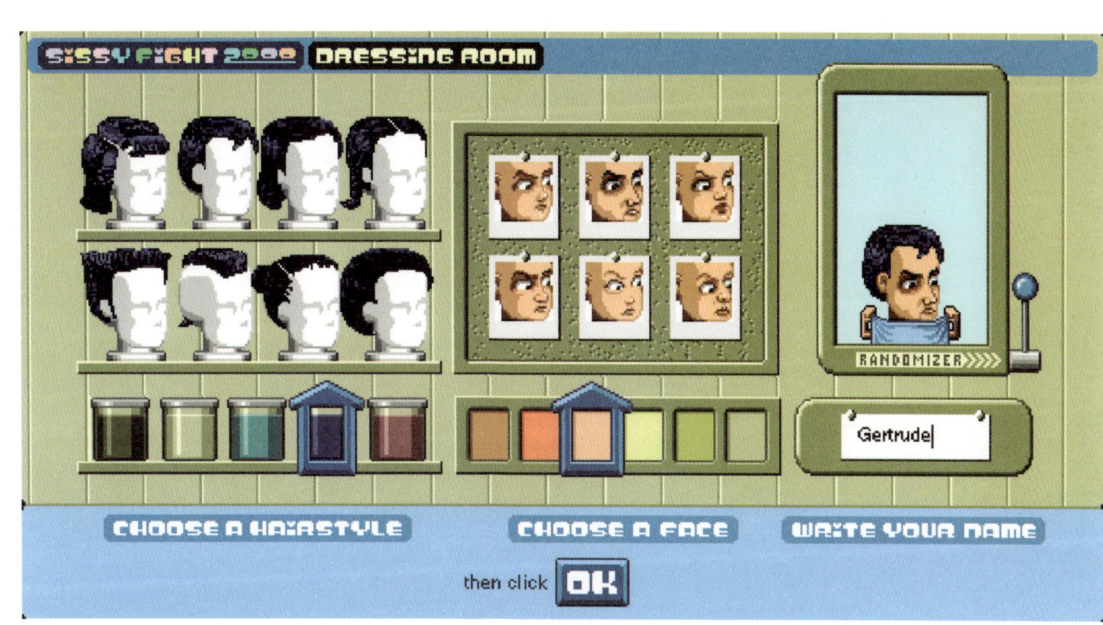

The prototype is tested, revisions are made, and the project is tested once more. In this way, the project develops through an ongoing dialogue between the designers, the design, and the testing audience.

In the case of games, iterative design means playtesting. Throughout the entire process of design and development, your game is played. You play it. The rest of the development team plays it. Other people in the office play it. People *visiting* your office play it. You organize groups of testers that match your target audience. You have as many people as possible play the game. In each case, you observe them, ask them questions, digest their answers, adjust your design accordingly, and playtest again.

This iterative process of design is radically different than typical retail game development. More often than not, at the start of the design process for a computer or console title, a game designer will think up a finished concept and then write an exhaustive design document that outlines every possible aspect of the game in minute detail. Invariably, the final game never resembles the carefully conceived original. A more iterative design process, on the other hand, will not only streamline development resources, but will also result in a more robust and successful final product.

SiSSYFiGHT 2000

> *SiSSYFiGHT 2000 is a multiplayer online game in which players create a schoolgirl avatar and then vie with three to—six players for dominance of the playground. Each turn a player selects one of six actions to take, ranging from teasing and tattling to cowering and licking a lolly. The outcome of an action is dependent on other players' decisions, making for highly social gameplay. SiSSYFiGHT 2000 is also a robust online community. You can play the game at www.sissyfight.com.[1]*

In 1999, I was hired by Word.com[2] to help create their first game. We initially worked to identify the project's *play values*: the abstract principles that the game design would embody. The values we created included designing for a broad audience of non-gamers; a low technology barrier; a game that was easy to learn and play but deep and complex; gameplay that was intrinsically social; and finally, something that aligned with the ironic Word.com sensibility.

These play values guided a series of brainstorming sessions, interspersed with group play of computer and non-computer games. Eventually, a game concept emerged: little girls in social conflict on a playground. While every game embodies some kind of conflict, we were drawn towards modelling a conflict that we hadn't seen depicted previously in a game. Technology and production limitations meant that the game would be turn-based, although it could involve real-time chat.

Once these basic formal and conceptual questions had begun to be mapped, the initial prototype took shape. The first version of SiSSYFiGHT was played with Post-it Notes around a table. I designed a handful of basic actions for each player, and acting as the program, I 'processed' the actions each turn and reported results back to the players, keeping score on paper.

Designing a first prototype requires strategic thinking about how to most quickly implement a playable version that can begin to address the project's chief uncertainties in a meaningful way. Can you create a paper version of your digital game? Can you design a short version? Can you test the interaction pattern with just a handful of players?

In the iterative design process, the most detailed thinking you need is that which will get you to your next prototype. It is important to understand the big picture too. Just don't let your design get ahead of your iterative research. Keep your eye on the prize, but leave room for play, accepting that some of your assumptions will be wrong.

The project team continued to develop the paper prototype, seeking the balance between co-operation and competition that would become the heart of the final gameplay. We refined the base ruleset—the actions a player can take each turn and the outcomes that result. These rules were turned into a spec for the first digital prototype: a text-only version on Internet Relay Chat (IRC), which we played taking turns sitting at the same computer. Constructing that early, text-only prototype allowed us to focus on the complexities of the game logic without worrying about other aspects of the game.

While we tested gameplay via the text-only iteration, programming for the final version began in Macromedia Director, and the core game logic we'd developed for the IRC prototype was recycled into Director code with little alteration. Parallel to this, visual designers began to develop the game's graphic language and chart possible layouts. Early drafts of the visuals (revised many times over the course of the project) were dropped into the Director version, and the first rough-hewn iteration of SiSSYFiGHT as a multiplayer online game took shape.

As soon as the web version was playable, the development team played it. And as it grew more refined, other Word.com staff were roped into testing as well. As the game grew more stable, we descended on our friends' dot-com companies after work, and letting them play cold. All of this testing and feedback helped us refine game logic, visual aesthetics, and interface. The biggest challenge was clearly articulating the relationship between player action and game outcome: because the results of every turn are interdependent on each player's actions, early versions of the game felt frustratingly arbitrary. Only through design revisions and tester dialogue did we manage to structure the results of each turn to unambiguously communicate what had happened that round and why.

When server infrastructure was completed, we launched the game to an invite-only beta-tester community that slowly grew in the weeks leading up to public release. Certain time slots were scheduled as official testing events, but our beta users could come online anytime and play. We made it very easy for the beta testers to contact us and email in bug reports.

Even with this small sample of a few dozen participants, larger play patterns emerged that needed to be tweaked. When the game did launch, our loyal beta testers became the core of the game community, easing new players into the game's social space.

In the case of SiSSYFiGHT 2000, the testing and prototyping cycle of iterative design was successful because at each stage we clarified exactly what we wanted to test and how. We used written and online questionnaires. We debriefed after each testing session. And we strategized about how each version of the game would incorporate the visual, audio, game design, and technical elements of the previous versions, while also laying a foundation for the final form of the experience.

Loop

Loop is a single-player game in which the player uses the mouse to catch flittering, coloured butterflies. The player draws loops around groups of butterflies of the same colour, or of groups in which each butterfly is a different colour (the more butterflies in a loop, the more points). To finish a level, the player must capture a certain number of butterflies before the sun sets. The game includes three species of butterflies and a variety of hazardous bugs, all with different behaviours. Loop was created by gameLab nd is available for play at Shockwave.com.[3]

Initial prototypes are usually ugly. They do not emphasize aesthetics or narrative content: they emphasize the game rules, which manifest as the internal logic of the game tied to the player's interaction. Visuals, audio, and story are important aspects, but the core uncertainties of game design, the questions that a prototype should address, lie in the more fundamental elements of rules and play.

Another way of framing this problem is to ask, "What is the *activity* of the game?' Rather than asking what the game is *about*, ask what the player is *actually* doing. Virtually all games have a *core mechanic*, an action or set of actions that players repeat as they move through the designed system of a game. The prototype should help you understand what this core mechanic is and how it becomes meaningful. Asking questions about your core mechanic can guide the creation of your first prototype, as well as successive iterations. Ideally, initial prototypes model this core mechanic and begin to test it through play.

Loop grew out of a desire at gameLab to invent a new core mechanic. There are ultimately few ways to interact with a computer game: players can express themselves through the mouse and keyboard, and games can express themselves through the screen and speakers. Deciding to intervene on the level of player input, we cast aside point-and-click or click-and-drag mouse interaction in favour of sweeping, fluid gestures.

The first prototype tested only this core interaction, allowing players to draw lines, but nothing else. Our next step was to have the program detect a closed loop and add objects that would shrink and disappear when caught in a loop.

Each of these prototypes had parameters adjustable by the person playing the game. The length of line and detail on the curve, the number of objects, their speed and behavior, and several other variables could be

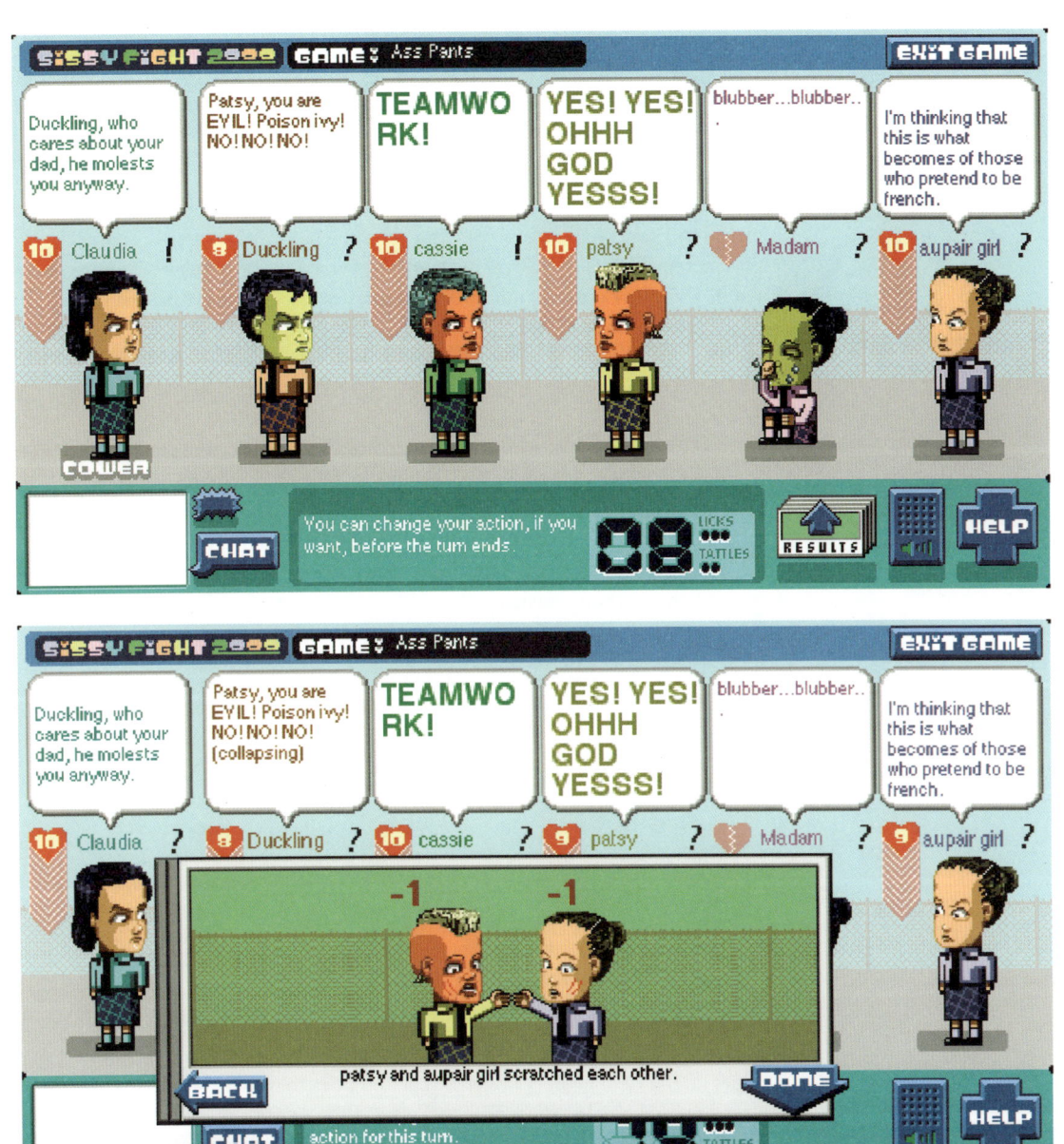

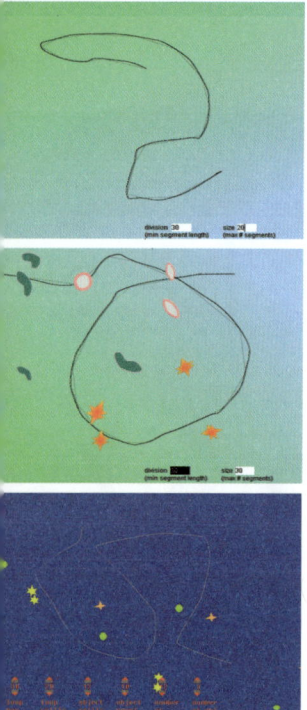

6-8 **Loop**
early prototypes

tweaked. As we played, we could try out parameters and immediately see how they affected the experience, adjusting the rules to arrive at a different sort of play. Building accessible game design tools into a game prototype is a technical strategy that incorporates and facilitates iterative design.[4]

As the butterfly content of the game emerged, so did debate about the game's overall structure and victory and loss conditions. Did the entire screen need to be cleared of butterflies or did the player just have to catch a certain number? Did the butterflies gradually fill up the screen or did their number remain constant? Was there some kind of time-pressure element? Were there discreet levels or did the game just go on until the loss conditions were met? These fundamental questions, which grew out of our core mechanic prototyping, were only answered by actually trying out possibilities and coming to conclusions through play.

As the game code solidified, the adjustable parameters of the game were placed in a text file that was read into the application when it ran. These parameters controlled everything from the behaviour of game creatures to points scored for different numbers of butterflies in a loop to the progression of the game's escalating difficulty. Thus game designers could focus on refining variables and designing levels, while the rest of the program—screen transitions, help functionality, high-score system and host-site integration—was under construction.

Loop's testing pattern was similar to SiSSYFiGHT's, moving outward from game creators to a larger circle of players. During development, gameLab created the gameLab Rats, our official playtesting 'club,' to facilitate testing and feedback. In the end, Loop achieved the fluid interaction we'd envisioned, an entire game evolving from a simple idea about mouse control. That is the power of iterative design.

LEGO Junkbot

LEGO Junkbot is a single-player game in which the player helps the robot character Junkbot empty trash cans throughout a factory. The player doesn't control Junkbot directly but instead uses the mouse to move LEGO bricks around the screen, deconstructing and reconstructing his environment brick by brick, building stairways and bridges that help Junkbot get where he needs to go. A variety of helpful and hazardous objects and robots add variety and complication to the game's sixty levels. Junkbot levels can be solved in multiple ways and the game structure encourages players to return to previously solved levels and complete them differently.[5]

The conceptual starting point for LEGO Junkbot came from gameLab's client, LEGO.com. LEGO wanted a game about brick construction with a target audience of eight- to twelve-year-old boys that that could also be played and enjoyed by adults. The challenge of the design was that real-world LEGO play was the referent. Yet we could never hope to recreate the sublime interactivity of plastic LEGO bricks. How could we translate LEGO play into a digital game?

facing page
9-10 **Loop**
final version

LOOP

"My pleasures are the most intense known to man: writing and butterfly hunting"

— *Vladimir Nabokov*

Loop me to
Play

14

11 **Loop**
final version

Our first step was to play with a LEGO bricks as a way of analyzing and understanding their subtleties. Then, we identified the project's play values. These values embodied the material and experiential qualities of LEGO as well as the cultural ethos of the LEGO play philosophy: modularity, open-ended construction, design creativity, multiple-solution problem-solving, imaginative play, and engineering. Using these play values, we brainstormed several game concepts.

The concept LEGO selected was called LEGOman (the character and storyline of Junkbot had not yet emerged) and it centred around moving bricks to indirectly help a character move through an environment. The first prototype was the simplest possible iteration thereof: players could use the mouse to drag bricks on the screen; there was a single, autonomously moving protagonist character; there were goal flags to touch; and there were rolling wheel hazards to avoid.

That first prototype was not very fun. When trying to invent new forms of gameplay, this happens sometimes. At this early juncture in the iterative design process, we could've scrapped the design and started fresh, building on insights from the unsuccessful prototype, or we could dig in and push on through. We chose the latter. Gradually we added elements, refining interaction, expanding level possibilities, and putting in new kinds of special bricks and robot hazards.

Each new element addressed something lacking in the experience of the previous prototype: It was monotonous to move bricks one by one, so we implemented code that let players move them as a group. We needed a way to move the main character vertically, so we added bricks which float Junkbot upwards. The game obstacles felt too deterministic, so we introduced robot hazards that responded in real time. As these interactive embellishments deepened the game (which was becoming fun), the character and storyline of Junkbot emerged.

Throughout, we utilized a level editor, a visual design tool that let game designers create and save levels. The editor allowed them to experiment with game elements and level designs, refining the overall experience and planning features for the next iteration.

Playtesting continued with the gameLab Rats using a web-based form to collect and collate testing data about the difficulty and enjoyment of each level. However, our main concern was whether the basic brick-construction core mechanic would be understood by our target audience, so we visited an elementary school computer classroom, sat kids down in front of the game, and let them play cold. This testing was invaluable, and confirmed our fears: too many testers had trouble picking up basic game concepts, such as how to make a stairway for Junkbot. This testing influenced the design, and we slowed down the learning curve, designing the first game levels to clearly communicate essential interactive ideas.

A good rule for iterative testing is to err on the side of observation. While it may be difficult to keep your hands off the tester's mouse, sit back and see what your audience actually does, rather than telling them

facing page
12 **LEGO Junkbot**
early prototype

13 **LEGO Junkbot**
final version

PARTICIPATORY CULTURE, DESIGN, AND ETHNOGRAPHY

LECO MAN

v0.7

EDIT **CATALOG** **PLAY**

JUNKBOT

PLAY ← CREDITS

WELCOME TO THE FACTORY
TRY MOVING THE COLORED BRICKS
WITH THE MOUSE
BEFORE YOU PLAY THE GAME

READY TO
PLAY

REPLAY INTRO

RESET SCREEN

how it's supposed to work. What you observe can be painful to watch, but it will help you design more successful play. Part of iterative design is simply learning how to listen.

Conclusions

Iterative design is a process-based design methodology, but it is also a form of design research. In each of these three case studies, new questions emerged out of the very process of design, questions that were not part of the initial investigation but that were nevertheless addressed through iterative play and design.

To design a game is to construct a set of rules. But the point of game design is not to have players experience rules—it is to have players experience *play*. Game design is therefore a second-order design problem, in which designers craft play, but only indirectly, through the systems of rules that game designers create. Play arises out of the rules as they are inhabited and enacted by players, creating emergent patterns of behaviour, sensation, social exchange, and meaning, thus the necessity of the iterative design process. The delicate interaction of rule and play is something too subtle and too complex to script out in advance, requiring the improvisational balancing that only testing and prototyping can provide.

The principles of the iterative process are clearly applicable beyond the limited domain of games. *Rules* and *play* are just game design terms for *structure* and *experience*: a designer creates some kind of *structure* (a typeface, a building, a car), and a reader, visitor, or car passenger *experiences* it, encountering, exploring, dwelling in, and manipulating the system—using it, playing with it, delighting in it. Games provide particularly clear examples of iterative design, but any design field can benefit from such an approach.

In iterative design, there is a blending of designer and user, creator and player. It is a process of design through the reinvention of play. Through iterative design, designers create systems and play with them. They become participants, but do so in order to critique their creations, to bend them, break them, and refashion them into something new. And in these procedures of investigation and experimentation, a special form of research takes place. The process of iteration, of design through play, is a way of discovering the answers to questions you didn't even know were there. And that makes it a powerful and important form of design research.

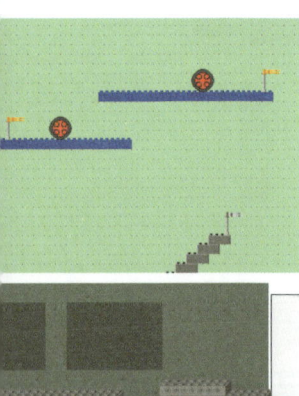

14 **LEGO Junkbot**
early prototype

15 **LEGO Junkbot**
level editor tool

Notes

This essay originally appeared in *Design Research: Methods and Perspectives*, edited by Brenda Laurel, and is reprinted as in its original text, without concession to standard Canadian spellings or grammatical forms, by request and kind permission of the author. Its inclusion acknowledges the important and lively contribution to the discussions on design methodologies and gaming, as applied to mobile technologies, made by Eric Zimmerman at the Mobile Nation conference and to this book.

[1] SiSSYFiGHT 2000 (Word.com, 2000).Project Team: Marisa Bowe, Ranjit Bhatnagar, Tomas Clarke, Michelle Golden, Lucas Gonze, Lem Jay Ignacio, Jason Mohr, Daron Murphy, Yoshi Sodeka, Wade Tinney, and Eric Zimmerman.

[2] In 1995, Word.com was founded as one of the web's first ezines. When the site folded, its domain name was sold to Merriam-Webster.

[3] Loop (gameLab, 2001) published by Shockwave.com.
Project Team: Ranjit Bhatnagar, Peter Lee, Frank Lantz, Eric Zimmerman, and Michael Sweet / Audiobrain.

[4] A sample of the game editor code:

```
-- LOOP SCORES
    score_same=0,5,10,20,40,80,150,250,350,500,700,1000,1400,1900,
    2500,3100,380
    0,4600,5500,7000
    score_different=0,0,30,75,200,500
    score_badloop=-20

    -- # of caught butterflies for each level of loop sound effect
    loop_sound_num=1,4,6,8,10

    -- BONUSES
    -- butterfly-borne bonus (x2):
    bonus_lifetime=60
    -- leaf-blown bonus (longer, moretime, freeze, flock):
    freebonus_speedlimit=15
    bonus_freeze_duration=4
    bonus_flock_duration=12

    -- HAZARDS
    snail_speedlimit=1.2
    killerbee_speedlimit=12, killerbee_attackrate=3,killerbee_stingduration=6
    beetle_speedlimit=3, beetle_fighttime=4, beetle_aborttime=10,
    beetle_effectradius=300
    stinkbug_speedlimit=2, stinkbug_tag_radius=40,
    stinkbug_effect_duration=10, stinkbug_effect_radius=300
    spider_speedlimit=9,
    spider_climblimit=22,spider_stingduration=6,spider_loop_length=5
```

[5] LEGO Junkbot (gameLab, 2001), published by LEGO.com.
Project Team: Ranjit Bhatnagar, Nick Fortugno, Peter Lee, Frank Lantz, Eric Zimmerman, and Michael Sweet / Audiobrain

Shaking Hands with the User
principles, protocols, and practices for user-integrated testing in mobile design

Barbara Crow and Kim Sawchuk
Evaluation, Mobility, and Usability and
Mobile Digital Commons Network

In designing non-commercial cultural experiences for mobile technologies such as cellphones, public participation is often invoked as a project goal. For most small design teams scrambling towards a deadline, doing wide-scale user testing of one's potential 'target markets' is not feasible. User testing can be expensive, time-consuming, perceived as a distraction from the 'real work' of designing, or seen as a form of project policing. Of course, not all designers share this point of view. At the Mobile Digital Commons Network (MDCN) we have been fortunate to work with designers who support user-integrated testing (UIT) in their design process. Developing feasible methods for small-scale UIT at the MDCN has been the central task of a working group known as Evaluation, Mobility, and Usability (EMU).[1]

EMU stands for the three conceptual pillars that uphold the MDCN evaluation component. It was created during the second iteration of the MDCN to gather more detailed information on how the imagined user might encounter and engage with mobile, wireless design projects. As one designer commented, he wanted more than the 'anecdotal evidence' that often comes with setting up a kiosk or an informal demonstration for friends or artists at a cyber-arts festival. This user integration is particularly urgent in projects that (a) are publicly funded, and (b) require individuals to deploy technologies, such as GPS, cellphones, or PDAs, whose habitual context of use is very specific. At the MDCN, artists and designers repurpose these familiar devices to present innovative, challenging cultural content.[2]

In this short report, we discuss EMU's methodological principles, how these principles have been translated into a concrete set of simple research protocols, and some of the benefits of integrating users into design for mobile platforms. These principles, protocols, and ideals (best practices) have been derived from conducting user evaluations in four MDCN projects: *Urban Archaeology: Sampling the Park, CitySpeak, Tracklines,* and *The Haunting.* While we draw heavily on standard research methods such as surveys, focus groups, one-on-one interviews, and participant observation, we call our approach *participant-creation* to signify the dialogic dimensions of our iterative integration of the wisdom of users *and* designers into the creativity of the research process.

[1] **Members of EMU 2004–2007:** Neil Barratt, Barbara Crow, Anna Friz, Alison Harvey, Ganaele Langlois, Andrea Zeffiro, Janice Leung, Kim Sawchuk

[2] **Visit our website** www.mdcn.ca/tiki-index. php?page=EMU and our publication *Wi* at www.wi-not.ca

facing page
[1] **MDCN workshop** Banff National Park

EMU Principles

One of our tasks as research designers has been to develop location-based research commensurate with the principles of locative media. For us, locative media involves attention to place (*loca*) and a sensitivity to how subjects move through a place (*loco*). As EMU, we try to develop specific research protocols appropriate to each project in connection with the research team which are built on our set of general principles and protocols. Codification of this process is crucial. We are committed to exchanging skills with designers so that they can implement their own field trials on location and put us out of a job. Our second principle is to commit to the concept of user integration rather than that of user testing. User testing, à la marketing research, treats the public as potential consumers who are a fount of information to be 'data mined.' Finally, in our 'search for a method,' to cite Jean-Paul Sartre's famous tome, we understand our shift, made via such classic techniques as participant observation or participatory design, as a move to participant-creation, which is principle number three.

These working principles have been derived from discussions with each other, and they draw upon feminist and post-colonial critiques of ethnographic research. However, it is not enough to have a set of principles. In UIT it is the subtle details and their implementation in situ—whether on the trail or in a public park—that matters. What follows is our checklist of UIT protocols that help us put our general principles into practice.

EMU User-Integrating Testing Protocols

1. Work closely with the design team to discuss and develop the field trial in relationship to the project. They may act as a member of the team.

2. Work with the team to develop a list of research questions and interview questions to ask your users. Keep the list short and work from the general to the specific in your questions.

3. Discuss observational strategies appropriate to the content of the piece and the place, and that are respectful to the users donating their precious time to the evaluation.

4. Decide what technologies you will use in the field trial, and double-check that they are working.

5. Find appropriate participants. Set a date. Stick to it, and be on time.

6. Make sure the necessary ethics forms and consent forms are ready.

7. Do a practice run with the research team before involving participants. It is recommended that the research team take the role of the research subject.

8. Conduct field test based on agreed-upon protocols set out by the team. Particularly important is this: Make sure that you keep your distance. Never let your desire for documentation lead you to crowd your subjects.

9. Debrief team after every test or day of test, if appropriate.

2 **Hoodoos Trail**
Banff National Park

facing page
3 **MDCN workshop**
Mount Royal Park, Montreal

PARTICIPATORY CULTURE, DESIGN, AND ETHNOGRAPHY

10. Do a major debrief after the entire trial is over to discern strategies for data archiving and analysis.

11. Begin archiving all documentation pertaining to the trial.

12. Analyze the data from memory and primary observations. This will guide your use of transcribed material, act as a synthesis of longer transcripts, and be used to verify differences between what is recalled and what was said.

13. Go back to the design team with observations and recommendations.

14. Permit the design team to work with this material and decide what is significant or not for the next phase of their work.

15. Use the experience to feed the design process and to set up, if desired, the next set of tests.

16. Consider, as applicable, making documentation available to participants. A content management system such as TikiWiki can facilitate communication amongst team members.

4 Banff National Park

EMU Best Practices: The Five Ds of User-Integrated testing

In our field trials with different MDCN projects we have noticed five contributions that a well-conducted user-integrated test can make to the design process. For the sake of memory, and alliteration, we name them the five Ds of user-integrated testing:

1. User-integrated testing is one way to make production *deadlines* happen.

2. Implemented at key strategic moments in the design process, UIT has the potential to provide valuable research *data* about the public that a designer may only vaguely imagine.

3. This data can act as a heuristic device leading to deliberation and assisting in decision making, thereby leading to *direction* in design.

4. Implemented with care, the ritualistic process and the deliberations that follow UIT can foster the *dedication* of the team to the overall project and create an effective sense of belonging that enhances collaboration.

5. Finally, UIT may produce valuable *documentation* that can be used for future projects.

These are, of course, ideal scenarios. Our version of UIT has not always been perfect, or yielded verifiable results that could be generalized into some grand statement or overarching theory about mobile users. Instead, valuable bits of information germane to particular projects have been gleaned: the desire of users for a map and a backstory prior to being sent out with a technology (*Urban Archaeology*); the desire of users for greater customization, as well as their concern with SMS costs, and their lack of familiarity with SMS in the Canadian context (*CitySpeak*); the default reversion of users to past media, such as audio guides, when using new media (*Urban Archaeology*); the importance of clear technical directions (*The Haunting*); and the way that temperature influences the movements of users and their willingness to listen to a lengthy narrative (*Tracklines*). Just as important are the methodological lessons we have learned: each field trial has been an ongoing apprenticeship in learning from falling and from failure as much as it has been a success in information gathering.

Locative media demands location-based research that is creative, flexible, open, and respectful of the ethical and epistemological conditions of research. It requires the ability to respond to accidents and the unexpected because we are not working in controlled experimental conditions: literally, we are often on the move walking beside and behind our subjects. However, when UIT is 'done right' it is a creative process of inquiry and communal knowledge construction that may imaginatively anticipate (rather than predict) a potential range of user responses to one's design goals. As one sound designer, who typically works for months in relative isolation, exclaimed with surprise after his first UIT field trial, 'It feels like you are shaking hands with the user!'

Informing Design through Ethnography and Informances

Ron Wakkary
Simon Fraser University Surrey

Introduction

Informance is a design technique that calls on designers to role-play and perform design ideas. It combines performance, role-playing, and improvisation, factors which enable designers to enact situations physically as well as conceptually.[1] We found that informance design used in design ethnography bridges ethnographic observations with the generation of design ideas. This approach emphasizes physical and embodied observations that are often overlooked, and guides formal analysis in the later stages of the process. These aspects are important. Firstly, because of the typical length of ethnography studies, designers need to be ready to deliver preliminary outcomes during ongoing studies. Secondly, ethnographic analysis tends to be textual rather than embodied and perceptual. This paper illustrates the value of informance in design ethnography by describing the process by which our observations led to design patterns and scenarios. In our case, we discuss a scenario based on a six-month ethnography study of everyday design.

Call-home Informance and Training the Mobile Phone Pattern

Leah, a design ethnographer, is enacting an observation of an informant, Kevin. Kevin is trying to use the voice recognition feature of his new mobile phone. Leah is in fact relaying how Kevin wanted to show her this feature. In the informance, Leah plays the role of Kevin and uses an eyeglasses case as a prop for the mobile phone. She imagines a Bluetooth headset is in place around her ear.

The informance starts with Leah acting as *Kevin* asking *Leah* if she wants to see how the voice recognition works: 'This is my mobile phone and this is Bluetooth…' *Kevin* puts the headset around his ear. 'It works best for me because I like to talk on the phone while I drive. I'd like to show you how it works.' 'Sure,' replies *Leah*. After punching keys on the mobile phone, *Kevin* holds onto the Bluetooth headset and says, 'Call home.' Nothing happens—Leah steps out of character to make some idle mobile phone sounds. Back in character he mutters to himself, 'Okay, it doesn't work,' then he says to *Leah*, 'sometimes it doesn't recognize certain syllables… let's try again.' Again Leah steps out of character to make phone sounds.

facing page
1 **Leah** performing Call Home informance

She resumes her role, holds the headset again, and says, 'Call home.' Again nothing happens. Furious punching of keys follows and then with some resignation *Kevin* admits, 'There are too many things on this mobile phone...' *Kevin* continues to push buttons. 'See, you know, certain words it works. I call CleanTech—here, I'll call CleanTech. CleanTech!' Triumphantly *Kevin* shows the display of the phone around the room. 'See it's connecting... it's connecting. So why won't it call home?' The informance ends as *Kevin* continues to push buttons. The informance lasted less than a minute.

In Leah's observations, *Kevin* never gave up on the voice recognition feature. The informance showed the balance between frustration, satisfaction, and perseverance with using a new feature when a need is identified as *Kevin* remarked about the use of his mobile phone while he drives. The observation behind this informance was later formalized in a pattern named Training the Mobile Phone (*fig.2*). Pattern language is a method design ethnographers have used to formalize observations.[2] Patterns describe existing routines and artifacts as design problems and solutions that have typically emerged over time.[3] The significance of the pattern Training the Mobile Phone was the observation that people will invest time and effort in integrating a new artifact if they see a long-term benefit that outweighs the short-term investment of trial and error. The informance aided the ethnographer's account of her recent observations.

Scenario: Playing Tag on a Whiteboard

In the middle of our study we were asked by a potential industry research sponsor for future scenarios of use of technology in the home. This request resulted in scenarios produced in a matter of days based upon the ethnographic data and patterns. One example is of an integrated whiteboard system that incorporates learning and testing by use on the part of the family. We excerpt part of the scenario here. The reader will find that the scenario is informed by the pattern and informance discussed in the previous section.

> *It's 7:30 a.m. Tuesday morning. Ken heads downstairs with a sense of urgency; as he's doing his tie, he's trying to call the office through his Bluetooth mobile-phone headset: 'Office. Office. Office.' It's new, and he is still training it. He pauses for a second on the stairs to grab the mobile phone out of his pocket. He begins cycling through the menu, attempting to debug the voice-activation feature. Remembering that words with sharp consonants seem to work better, he tries to set up a different cue: 'Work. Work. Work.' It's not working. He's been trying to fine-tune this system for a couple of weeks now, knowing that once it's bug-free it will save more time than dialing by hand, and free him to make calls while driving. He'll give it another shot tomorrow; right now, he's got to get on with his day.*
>
> *Before heading out the door, he visits the kitchen to grab a bagel and coffee; as he reaches for the fridge, he notices that Madison has written a note saying 'Soccer @ 7:00pm.' He thinks it's a reminder that he has to drive her. Ken checks his mobile phone again to double-check his schedule for*

2 **Training the Mobile Phone**

Significance: People will invest time and effort in integrating a new artifact if they see a long-term benefit that outweighs the short-term investment of trial and error.

Description: Changing our routines, or taking the time to train others or learn new systems in order to incorporate a new artifact or system.

Example: Kevin made repeated attempts to train his mobile phone voice recognition because he knew it would save him time in the future.

Related Patterns:
• Testing a Chalkboard
• Wiping the Dog's Paws

the day. Looking at today's date, he sees the soccer practice reminder listed in a yellow box; he taps the box with his finger and the checkmark turns to green. 'You're lucky the mobile phone is smart about some things, Madison,' he thinks as he chuckles to himself. This time, Madison correctly tagged the message for Ken and it was in fact the whiteboard that sent the message to Ken's mobile phone...

Conclusion

Design ethnography is a powerful analytical approach for design that relies on representations leading to design outcomes. We have shown that informances bridge the gap between observations and design ideas when used as an analytical tool in sequence with patterns and scenarios.

Acknowledgements

We are grateful to the families who participated in our study and opened their homes to us. We acknowledge the support of this research through a grant from the Social Sciences and Humanities Research Council of Canada.

Notes

[1] B. Laurel, 'Demo: Design improvisation,' in *Design Research: Methods and Perspectives*, ed. B. Laurel (Cambridge: MIT Press, 2003) 49–54.

[2] A. Crabtree, *Designing Collaborative Systems: A Practical Guide to Ethnography* (London: Springer, 2003); D. Martin, T. Rodden, M. Rouncefield, I. Sommerville, and S. Viller, 'Finding patterns in the fieldwork,' in *Proceedings of the Seventh European Conference on Computer-supported Cooperative Work* (Bonn: Kluwer Academic Publishers, 2001) 39–58.

[3] C. Alexander, S. Ishikawa, and M. Silverstein, *A Pattern Language: Towns, Buildings, Construction* (New York: Oxford, 1977).

Mobile Phone Imaging as Gesture and Momento

Rob Shields
University of Alberta

Mobile phone images and videos extend the genealogy of popular camera technologies in ways which disregard concerns over resolution, quality, and robustness in favour of the production of digital glances which are props and momentos[1] testifying to photographers' relation to their subjects—'Look! It's us five seconds ago!' Although they can be used for snapshots, although they could be a useful technology for visualizing the city, mobile phone imaging also witnesses ourselves back to us. This social reflexivity emphasizes relationality and does so with little time delay, allowing photographing to become an activity embedded within everyday social interaction. This raises many questions concerning what exactly is made 'visible' in the gesture of mobile phone imaging. For example, How does it change what we understand to be legitimately exposed as 'public' versus hidden and 'private'? How does the process of making mobile images and clips shift the way things are 'brought to the stage' or pass 'under our noses' in everyday life? This may include the virtual or intangible aspects of urban life, changing the way in which we understand social interaction and the balance between the elements that go into making up what we understand as the 'reality' of urban situations. The question of staging is also a matter of power, of the making and remaking of 'normality.' A video exercise for the Bahia Celular Filme Festival de Cinema Mini-metragem illustrates this problem from the point of view of a foreign tourist confronting the *Orixás*, the hybrid gods of the city of Salvador, Brazil's black Vatican of Candomblé.

The now-common ability to take photographs with mobile phones brings photography into daily life in a way which once characterized only professionals and amateurs who carried a camera everywhere. The weight, investment, and obviousness of a camera meant that even if one did not identify oneself as a 'photographer' or photography buff, one quickly became thought of as such. The twentieth-century history of photography has been characterized by the search for unobtrusive portable cameras with lenses which maintain an acceptable visual detail and offer high enough shutter speeds to capture fleeting events. A historical trajectory thus links 35 mm cameras, the SLR, and convenient zoom lenses to small digital cameras and the development of mobile phones with improving picture-taking features.

[1] **Momentos**, which are distinct from more commonly discussed *mementos*, are defined within this paper.

facing page
1 **Gesture**
The phone as a prop: interaction with immediate images uniquely structures the making of mobile phone images

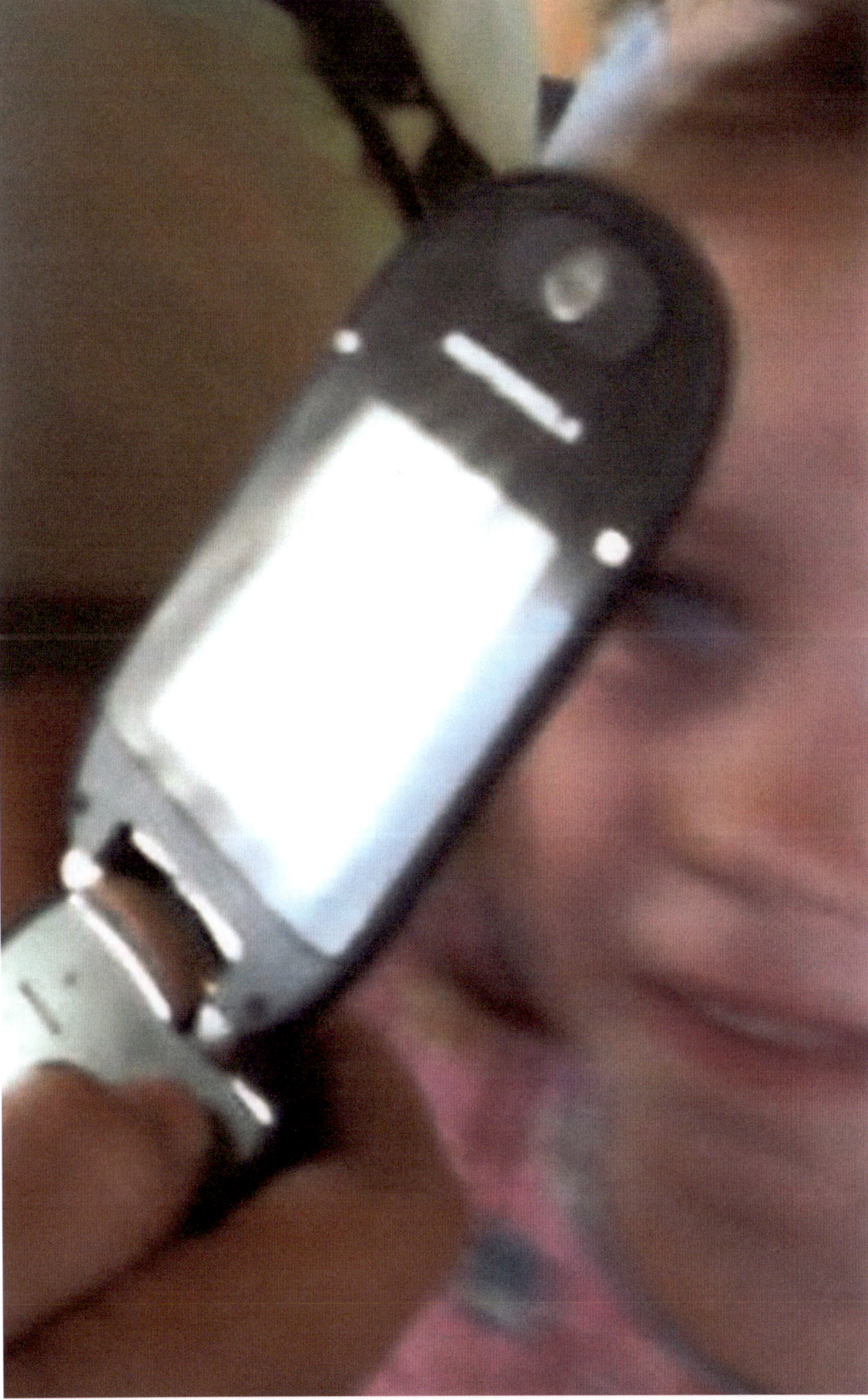

The distinction between the cameras built into mobile phones and dedicated cameras lies in the quality of their lenses and imaging abilities. Most mobile phones remain under two megapixels in image resolution. This is sufficient for a 'thumbnail' or even a two-inch by three-inch image, but incomparable with the expectations of either digital photographs viewed on computer screens or camera users who expect snapshots of resolution sufficient to produce a quality image at a minimum size of four inches by six inches. They cannot be used for professional imaging. Nor do they capture sufficient detail in a wide enough range of lighting conditions to be used as an adjunct to professional activities, for example to document situations where liability or capital is at stake (for example, architects noting the progress of construction or real estate agents documenting properties for sale).

For example, mobile phone cameras are notable in being the province of manufacturers whose primary focus is electronics—they do not appeal to consumers by attempting to borrow the brand identities of major camera manufacturers. Nor do they advertise the quality of their optics.

How did resolution become immaterial to mobile phone cameras? In contrast with cameras, with mobile phones the emphasis is heavily on taking pictures of people and of sites 'on impulse.' The process of taking a picture is constructed as a joint activity linking mobile phone photographers and their subjects. The device itself comes to be seen as a medium which links and binds the interactants, rather than as a technology which primarily links 'real life' to an archival image. Pictures can be immediately viewed on the phones' screens as a further joint activity of all involved and sent to others' mobile phones to be collected for later viewing on their phones. This is a digital form of the practice of carrying wallet-sized photos of loved ones. Like all photographs, images recalled on mobile phone screens prolong and supplement memories of occasions and places. Reflexively, images of oneself that a person identifies with can become digital versions of cartes de visite produced by nineteenth-century photographic studies, as can be viewed in the McCord Museum's extensive William Notman collection.

This process-based emphasis is both the construction of mobile phone manufacturers who market the 'fun' of mobile phone photography rather than its intrusiveness or awkwardness. In addition, cameras are presented in promotional materials as devices which seem to be at the centre of groups of people, rather than as the sort of objects which rarely leave their owners hands or pockets. They are used as reflexive technologies whose owners can contribute images to group interactions creating a type of social reflexivity.

I do not deny that mobile phone cameras are also used like 'instant cameras' such as the Polaroid film cameras to which mobile phone cameras are digital and miniaturized cousins. Nor do I want to suggest that these devices are not used to take snapshots. Their popularity, the concern over their usability, and the lack of attention to resolution relates them more closely to the 'Kodak philosophy' of popular photography than even to

facing page
2 **Still from mobile phone video *Orixas*** Bahia Celular Filme Festival de Cinema Mini-metragem 2006

PARTICIPATORY CULTURE, DESIGN, AND ETHNOGRAPHY

2 **Roland Barthes,**
Camera Lucida (New York:
Hill and Wang, 1981), 5.

professional digital cameras (large-formats and SLRs). Even though Barthes saw that 'The Photograph is never anything but an antiphon of "Look," "See" and "Here it is"; it points a finger at certain vis-à-vis, and cannot escape this pure deictic language,'[2] there is something added or extended in the case of mobile phone cameras.

In comparison, then, to most other cameras, the use of the mobile phone camera involves a different performance, and its digital images are complemented by a different spectatorship which looks for the punctum rather than the studium (or content) of images. The 'question' or problem of the mobile phone image is much more strongly not one of literal content but of what is/was passing through the moment captured in the time of the image. The time-image and movement-image do not coincide. Although the moment captured is past and, as one says, 'gone' (where?), the movement or flow of time and events which passed through the moment continues into the present of the spectator. In this way, an understanding of mobile phone images seems to be less dominated by the questions of death and irretrievability of the past which preoccupy Barthes, whose central example is a photograph of his dead mother. The image blurs the distinction between a present spectator alive to the image, and the liveliness of the past moment captured by the image—its virtuality.

Mobile phones could be the basis of a nascent archive of intimate and informal occasions and passing moments: personal visual souvenirs. One could thus hope that mobile phone photography offers a new vision of the city. As a visualization technology and as an art form, mobile phone imaging may also yield exhibition material. In newsworthy events, their availability offers a way of visually recording eyewitness experiences. Consider the scandal of the photographs of the execution of Saddam Hussein.[3] However, I suspect most mobile phone photographs and videos will not be archived; they will be lost to the future, like an oral art.

3 **For a discussion** of the
implications of this particular
cellphone video, see Rob
Shields, 'Saddam Hussein
Video: We too Were There,'
Space and Culture blog,
posted on January 6, 2007,
www.spaceandculture.
org/2007/01/saddam-hussain-
video-we-too-were-there.php.

The tendency is to treat mobile imaging as merely a new visualization tool; however, there is a risk that we will overlook the new ways in which mobile phone cameras 'make visible' both the usual subjects of photography and new aspects of situations. This is very different from the abstract sense of the mobile phone as an interface into and platform for a personalized and branded information space. They are not simply 'representing' or imaging the city in a new way. Rather they are shifting how the city is *imagined*, the way in which everyday life is caught up not with representing space but 'in a space of representation.'[4] However, it complements the notion that mobile devices speed up information flows to make suburban sprawl and urban gridlock manageable and livable for individuals dealing with the increasing scale of cities. Having to move around major cities and connect with others—for meetings or to locate them—is unimaginable if one is continually delayed due to congestion or confused in unfamiliar neighbourhoods.

4 **Henri Lefebvre,**
The Production of Space
(Oxford: Basil Blackwell,
1991).

How do the imaging and video capabilities affect how users relate to cities, find their ways through them, or coordinate their interactions with others? How does video in particular fit into the process of 'tuning' and updating one's information with continual short calls and interruptions from others reporting their progress in everyday life? Are these different from the ways which have become doxa in developers', politicians', and urban planners' understandings of cities?

Mobile phone photography actualizes an interactive medium between photographers and subjects. Their aesthetics are not those of formal graphic composition but of the presence of a detail which crystallizes a moment more extensive in time, or a scene which affirms the activity or experiences of the photographer. It does not follow professionalized norms of lighting, resolution, and focus. Mobile phone photography can be understood as a form of 'witnessing' which links actual events in one space and time with a longer duration than the instant a shutter opens and closes (the virtual elements of an event). It can be a momento:

+ A relation between photographer and subject.
+ A process (taking pictures of the city on a rare walk home, coming into a new relation with it).
+ A representation of an ongoing social interaction as a type of prop.
+ A means of reconnecting a viewer with a past event as an *aide-mémoire*.

Here the emphasis is on the 'doing,' the making of a photographic relation. Because an intangible relation is made visible, it changes the weight we give to such virtualities. 'Spontaneous' images and videos subtly intervene in how we understand the reality of a moment. Although subtle, this is a change in our world and in our cities. What passes through the mobile photographic and cinematographic glance, like the momentary flicker of alarm on a subject's face, acquires the air of an authorized version of the moment. Affects that do not pass through the image captured in a particular moment slip away. Mobile phone images made by holding the phone at arm's length and turning it on oneself and one's group or party further amplifies this memorial quality and the sense of the phone as an 'actant' or prop participating in and affecting the ongoing interaction in ways which may not be obvious to the photographer, the spectator(s), or the participants.[5]

5 **Bruno Latour,** *Pandora's Hope* (Cambridge: Harvard University Press, 1999).

Acknowledgements

Thanks to the commentators on previous versions of parts of this paper which appeared on the Space and Culture blog at www.spaceandculture.org/2006_10_01_archive.php

Being There

uncanny medium, methodological multiplicity and proliferative embodied creativity in *The Haunting*

David McIntosh
Ontario College of Art & Design

This paper examines the processes that proliferate creative autonomy and agency through the networked boundary object medium of the cellphone, through a methodological multiplicity that draws on divergent art and design methods, and through the intense, embodied production method of the charette, all applied in the iterative design process underpinning *The Haunting*. The paper opposes theoretical and methodological purity, and rejects claims for a new, scientized 'real.' The motivation for this intensive, proliferative methodological approach anticipates and speaks to the imminent conversion of cellphone users globally into both producers and consumers of games as more sophisticated phones, user-friendly game engines, and horizontal communications technologies proliferate. Before addressing the specifics of the methods employed in the two-year production of this cellphone game, I want to establish a broader theoretical and historical frame for my approach to constructing mobile locative interactive media games, beginning with a reconsideration of the cellphone as medium.

Telephonic Medium as Uncanny Thing: From Facts to Animism

A popular historical driver of new representational applications for new communications technologies has been communication between the realms of the living and the dead. As Erik Davis has points out in his book *Techgnosis,*

> the final decades of the nineteenth century were actually boom years for pop mysticism, occult science and decadent romanticism… Mesmerism, Spiritualism, Theosophy and Mary Baker Eddy's 'Christian Science' all expressed the desire to ensoul science, to overcome the growing divide between rationalism and religion… Daguerreotypes, phonographs, telegraphs, telephones—all these nineteenth-century media siphon a bit of soul into an artifact or an electric herald.[1]

This historical divide between the living and the spirits of the dead, between rationalism and religion, sutured by new information and communications technologies, echoes through the deployment of the cellphone medium in *The Haunting*.

facing page
1-2 *The Haunting*
visualization

3-4 **The Allan Sisters' Spirits**
performed by Jennie Ziemianin
and Nevena Niagolova

'Medium' is a complex and slippery term comprising a network of multiple and simultaneous meanings. In communications, medium refers to a technology and its related techniques; in the paranormal, a medium is an intermediary between the living and the dead; in painting, a medium binds the pigment to a surface; in clothing and latte terms, medium is a relative term for something in between large and small. In fact, the 'uncanny telephonic' medium grew out of a merging of spiritualism and science,[2] notably in the work of Alexander Graham Bell's assistant Thomas Watson. Bell developed the concept of converting of the human voice into an electrical signal to pass along a wire, but Watson actually built it. A member of the Society for Psychical Research, Watson treated spiritualism as a non-mystical science: 'Bell and Watson actually tested talking via phone by sending a weak current through a séance-like circuit made up of a dozen people holding hands.'[3] Over the years, animist understandings of the telephone as a sentient thing that sutures spirit and science have assumed many forms. Most recently, the cellphone has assumed mainstream mass media narrative importance as a sutured spirit/science fetish sign, especially in Asian societies where cellphones are ubiquitous. There has been a recent flurry of feature films produced in these societies, all based on the narrative device of the *telepon hantu*, or 'the ghost in the phone.'[4] Some titles in this horror sub-genre include *Cell Phone* (China, 2003); *Love Message* (China, 2005); *Phantom Call* (China, 2000); and *Samurai Cellular* (Japan, 2000). The most popular such production so far has been *Phone* (South Korea, 2002), remade in Japan as *One Missed Call* (2004) by the infamous Takashi Miike.[5] In many ways, *The Haunting* rests on a structuralist, reflexive turn in this history of mediating between the living and the dead and a similar turn in the popular construction of the cellphone as 'science-meets-spirit' fetish sign in that the cellphone is made to serve simultaneously as means of production, means of consumption, and narrative content.

The network of meanings attributable to the concept of medium in general, and to the cellphone in particular, reflects a number of theoretical approaches in understanding and representing complex phenomena and the perceived divide between facts and spirits. Bruno Latour suggests that there has been 'a huge sea change in our conceptions of science, our grasps of facts, our understanding of objectivity. For too long, objects have been wrongly portrayed as matters-of-fact. They are much more interesting, variegated, uncertain, complicated, far reaching, heterogeneous, risky, historical, local, material and networky than the pathetic version offered too long by philosophers.'[6] Paraphrasing Latour, perhaps objects are more rightly portrayed as 'matters-of-factish,' as hybrids of facts and fetishes.[7] Latour's concept of the 'matter-of-factish' boundary object is echoed in a constellation of theoretical tools for conceiving complex representational processes, ranging from Trinh T. Minh-ha's reflexive interval in documentary, where 'On one hand, truth is produced, induced, and extended according to the regime in power. On the other, truth lies in between all regimes of truth,'[8] through Michael Taussig's oscillatory mirror dance between mimesis and alterity where 'the flip-flop from spirit to thing and back again—the decided undecidability—[writes] incantatory spells of mimetic-realism, where the spirit of the matter meets the matter of the spirit' and Lev Manovich's impure new media interface.[9] It is in these spaces of flow where sedimented differential power relations begin to shift and recombine. In specific relation to the 'matter-of-factish' cellphone medium, these theoretical tools serve as conductors for the complex, proliferative dynamics between physical space and Hertzian space, between the living and the spirit world, between embodiment and virtuality.

Sampling Methodological Multiplicity and Innovative Intersections

With this theoretical modelling of networked boundary object dynamics in mind, I want to focus on developing a parallel constellation of art and design methods, all outside of the cellphone medium, that can assist in deploying methodological multiplicity as a force for proliferating creative outcomes and for enhancing self-representational autonomy and agency. Conventional understandings of art production suggest that methods revolve around spontaneous, experiential, provisional, metaphoric, heuristic, and analogic imaginings and improvisations where the prototype is arguably the art. Conventional understandings of design production suggest that methods are directed toward problem solving, task definition, and information visualization, where a predetermined set of rules, processes, and subroutines transform the prototype into final product. While these methodological distinctions may appear oppositional, through the theoretical tools of the 'matter-of-factish' boundary object, the reflexive interval, the mirror dance, and the impure interface, they can be reconceived and recombined as nodes in a much larger network of intersecting and interacting art and design methodologies. By examining the specific methods of several artists and designers, I want to model a sample network that demonstrates how methodological multiplicity reveals innovative intersections across what have been portrayed as methodological differences in art and design.

Jack Smith, renowned film and performance artist, maintained his works as unstable objects, re-editing his films as they went into the projector and reconstructing every iteration of his films and performances as a new prototype, as opposed to his contemporary and nemesis Andy Warhol, who made mass reproduction his artistic mission. In 'Capitalism of Lotusland,' Smith questions, 'Could art be useful? Ever since the desert glitter drifted over the burnt-out ruins of Plaster Lagoon, thousands of artists have pondered and dreamed of such a thing, yet, art must not be used anymore as another elaborate means of fleeing from thinking because of the multiplying amount of information each person needs to process in order to come to any kind of decision about what kind of planet one wants to live on before business, religion, and government succeed in blowing it out of the solar system.'[10] Smith's method of engaged, baroque, anarchic heuristics and eternal self-reinvention intersects in interesting ways with Mark Lombardi's vanguard schematic visualizations of corrupt political-economic networks, where the prototype becomes the art. As Lombardi describes his method, 'I originally intended to use the sketches solely as a guide to my writing and research but soon decided that this method of combining text and image in a single field really worked for me in other ways as well… Prior to that I felt like I was some kind of schizophrenic. I kept thinking I needed some type of medium or vehicle of my own devising that would unite the two, and now, with this schematic form, I had it.'[11] Noted art critic and historian T.J. Clark underscores the urge to self-representational agency through the methodological approach of the lyric self:

> It seems I cannot quite abandon the equation of Art with Lyric. Or rather—to shift from an expression of personal preference to a proposal about history—I do not believe that modernism can ever quite escape from such an equation. By 'lyric' I mean the illusion in an artwork of a singular voice or viewpoint, uninterrupted, absolute, a world of its own. I mean those metaphors of agency, mastery, and self-centeredness that enforce our acceptance of the work as the expression of a single subject.[12]

And contemporary artist Germaine Koh's method involves constructing new media works that are not just science-objects but networks of anonymous and evanescent relations, as in her 2006 installation *Call*.[13]

Working in an AI design environment, Balakrishnan Chandrasekaran takes a highly systematized approach to design method as problem solving in 'Design Problem Solving: A Task Analysis.' For him, design is a deliberative and synthetic process involving 'mappings from the space of design specifications to the space of devices or components, typically conducted by means of a search of possible subassemblies of components.'[14] He proposes the algorithmic Propose-Modify-Critique (PCM) method for task structure design. He acknowledges that there is another view of design method as 'an intuitive, almost instantaneous, process where a design solution comes to the mind of the designer' but claims the scientific 'real' with his method, suggesting that 'even when a plausible solution occurs in this way, the proposal still needs to be evaluated, critiqued, and modified… deliberative processes are still essential for real-world design.'[15] In *Information Visualization*, Colin Ware affirms and expands on Chandrasekaran's rigid approach, offering slightly more expansive improvisational methods at certain points. Ware's creative problem solving through information visualization breaks problems down into tasks—preparation, production, judgment—much as Chandrasekaran's method does. However, Ware accepts and promotes discovery and invention as crucial elements of the problem-solving interface, which must 'support the rapid creation of loose sketches, the ability to modify them, and the ability to discard all or some of them.'[16] For Ware, this interface must be simple enough to not inhibit the visual thinking process; for example, a constructive diagram done as a rapid sketch on a napkin promotes multiple interpretations and/or design hypotheses.

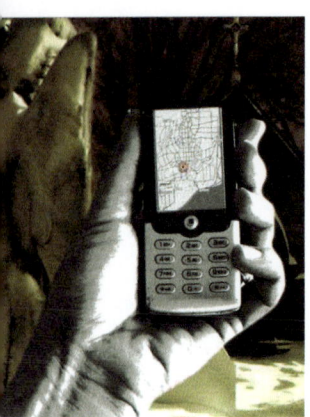

5 *The Haunting* visualization

PARTICIPATORY CULTURE, DESIGN, AND ETHNOGRAPHY

However, it is important to note that Frank Gehry sketches, proto-types, or hypotheses in his complex design and building process are now sold as art objects in their own right, a design phenomenon that takes us back to the art-world prototypes of Jack Smith and Mark Lombardi. Bernard Reber offers an expanded understanding of multiple methods applied to the design of publics and democratic processes through the socio-political experimentation of Participative Technological Evaluation (PTE).[17] He describes a methodological wealth for designing new publics, including

> citizen juries, consensus conferences, deliberative conferences, Delphi and charette methods, expert panels, focus groups, planning teams, script-writing workshops, consumer workshops, global cafes, direct initiatives and referendums, public surveys, public auditions, opinion polls, multiple choice questionnaires, discussion and negotiation between interest groups, citizens' councils, voting conferences, interactive technological evaluation, interdisciplinary working groups and political role-playing.[18]

This sampling of a range of distinct art and design methods in a number of media and disciplines reveals a rich network of intersections, contradictions, complexities, and proliferations produced through the superimposition of seemingly oppositional methods. And it is this meth-odological multiplicity that connects the theoretical model of 'matter-of-factish' boundary objects as sites for improvisational recombinance and power realignments to innovative reimaginings of the cellphone medium in *The Haunting*.

Charette as Improvisational Interval and Embodied Interface

The Haunting involved extended research and creation, over the course of two years, in two distant locations, a structural arrangement that would seem to go against the grain of designing a cellphone game located in one very specific site. The distributed, multiply located design and production team on *The Haunting*, with one team in Toronto at the Ontario College of Art & Design (OCAD) and the other team in Montreal at Concordia University, produced a unique set of methodological challenges that were addressed through the charette. The charette is another instance of a medium, boundary object, interval, and interface that was deployed in this instance to transform decentralized virtual work in a network into a series of embodied and intensive face-to-face sessions, where participating artists and designers, undergraduate students, and professionals brought their skills and methods to bear on the project. *The Haunting* developed from three four-day charettes—two at Concordia and one at OCAD. I supervised the first *Haunting* charette where a van of interns—under-graduate students from art and design faculties at OCAD—went to Montreal for four days of work with Concordia interns. *The Haunting* charettes served as reflexive bridges between the virtual/embodied divide in the production of the game, much as the cellphone medium sutures the spirit/science, presence/absence divides in the narrative

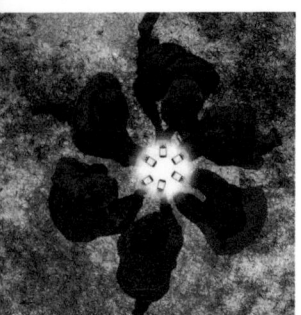

6 **The Haunting**
visualization

of the game. In addition to their crucial methodological role in contextualizing virtuality/embodiment, the charettes provoked new expressions of imagination and improvisation through their intensity. The charettes involved collective participatory design and rapid prototyping, where brainstorming, scripting, asset production, interaction engineering, and on-site testing were all accomplished in six-hour periods. The usual delay between concept, testing, and reiteration was drastically reduced to produce an even more intense dynamic across medium, mobility, site specificity, and virtuality where improvisational responses proliferated. Methodological multiplicity applied in the context of the medium-as-boundary-objects of cellphone and charette, spaces of flow where power relations are reconfigured, promoted the decomposition of the sedimented fetish sign of the cellphone medium by peeling the signifier off the signified and then rebuilding multiple meanings through improvisational recombinance. The following examples demonstrate how the various mediums and methods intersected to produce networks of new meanings.

The conceptual instigation for *The Haunting*, where the cellphone sign was initially de- and recomposed, occurred when I saw a group of four young men, walking together at night, all looking into their cellphone screens. The eerie blue glow that lit their faces provoked a re-metaphorization of the phone as light source and as a flame in a campfire around which ghost stories are told. The fact that each of the young men was looking into his own cellphone provoked further metaphoric improvisation, reframing each phone in the grouping as a part of a larger collective puzzle. These initial destabilizations and reassignments of meaning were subsequently developed in participatory design games, where the group solved puzzles and problems collectively, and in a cellphone circle where a moving image of fire was triggered by Bluetooth proximity to flash from one phone to the next, around the circle of assembled phone screens. During a field test on Mount Royal, one of the artist participants from OCAD used the bottom of a small pool of water in the woods as a screen on which he projected a moving image directly from the cellphone screen. This further improvisational deployment of cellphone as light source was then developed more formally into a moving image riddle of a spirit swimming up from the bottom of a dark pool of water that could only be solved by holding the image to a mirror. In turn, this underwater imagery connected directly with documentary research into the dead who were buried in cemeteries on Mount Royal, specifically, Gwen and Anna Allan, sisters and granddaughters of wealthy Montreal banker Hugh Allan. The sisters drowned in the sinking of the *Lusitania* in 1915; Anna's body was never found, while Gwen's body was buried in the family plot on Mount Royal. I conducted additional documentary research into the history of Arthur English (a.k.a. Arthur Ellis), Canada's last official hangman, who was buried on Mount Royal. The Allan sisters' and English's documentary histories have continued to develop as parts of the 'matter-of-factish' narrative of spirit possession on the mountain. In the process of developing visual content for the cellphone screen, the cellphone was reimagined by charette participants as a cage for these spirits on the loose, with the frame of the screen serving as the walls of the cage. I also re-metaphorized the cellphone as the marker on a ouija board through which direct contact between the living and the dead was made, a reimagining at the centre of many of the design visualizations of *The Haunting*. The intensive charette process outlined here exemplifies how a multiplicity of art and design methods can be deployed to generate a multidirectional and expanding network of improvisational meanings and interactions for a mobile locative experience.

Notes

[1] Erik Davis, *Techgnosis: Myth, Magic and Mysticism in the Age of Information* (New York, Three Rivers Press, 1998), 65.

[2] Davis, *Techgnosis*, 64.

[3] Davis, *Techgnosis*, 66.

[4] B. Barendregt, 'The Ghost in the Phone and other Tales of Indonesian Modernity,' in *Proceedings of the International Conference on Mobile Communication and Asian Modernities* (Hong Kong: City University of Hong Kong, 2005), 64–67.

[5] Manuel Castells, *Mobile Communication and Society: A Global Perspective* (Cambridge: MIT Press, 2007), 124.

[6] Bruno Latour, 'From Realpolitik to Dingpolitik: Or How to Make Things Public,' in *Making Things Public: Atmospheres of Democracy*, eds. Bruno Latour and Peter Weibel (Cambridge: MIT Press, 2005), 19–21.

[7] Bruno Latour, *Pandora's Hope: Essays on the Reality of Science Studies.* (Cambridge: Harvard University Press, 1999), 291.

[8] Trinh T. Minh-ha. 'The Totalizing Quest for Meaning,' in *Theorizing Documentary*, ed. Michael Renov (New York: Routledge, 1993), 90.

[9] Michael Taussig, *The Nervous System* (New York: Routledge, 1992), 10; Lev Manovich, *The Language of New Media* (Cambridge: MIT Press, 2001).

[10] Jack Smith, 'Capitalism of Lotusland,' in *Wait For Me at the Bottom of the Pool: The Writings of Jack Smith*, eds. J. Hoberman and E. Leffingwell. (London: Serpent's Tail, 1997), 11.

[11] Mark Hobbs, *Mark Lombardi: Global Networks* (New York: Independent Curators International, 2003), 16, 34.

[12] T.J. Clark, *Farewell to an Idea: Episodes from a History of Modernism* (New Haven: Yale University Press, 1999), 401.

[13] Germaine Koh, *Call*, www.germainekoh.com/call.html.

[14] B. Chandrasekaran, 'Design Problem Solving: A Task Analysis.' *AI Magazine* 11 no. 4 (Winter 1990): 60.

[15] Chandrasekaran, 'Design Problem Solving,' 60.

[16] Colin Ware, *Information Visualization: Perception for Design* (New York: Morgan Kaufmann, 2004), 385.

[17] Bernard Reber, 'Public Evaluation and New Rules for "Human Parks"' in *Making Things Public: Atmospheres of Democracy*, eds. Bruno Latour and Peter Weibel (Cambridge: MIT Press, 2005), 314.

[18] Reber, 'Public Evaluation,' 315.

The Haunting
voices from beyond in mobile experience design

Michael Longford
Concordia University

¹ **For more information** see
related essays in this volume
or www.thehaunting.ca.

The Haunting is a multiplayer, location-based cellphone game designed for Mount Royal Park in Montreal.[1] Players are invited by fictional provider VFB Mobility to use a cellphone as the latest in technological means for exploring paranormal disturbances and communicating with the dead. Media debris, flickering screens, unearthly vibrations, and screaming cellulars inhabit the 'forest of shadows' surrounding the cross at the summit of the mountain in the centre of the park. Interaction scenarios, alternative mapping techniques, spontaneous public performances, and location-based play structures rooted in non-linear narrative are explored in mobile experience design. Using Global Positioning System (GPS) and Bluetooth beacons in a networked environment, this project treats territory as interface, playing with the potential of mobile technologies to augment and enhance our relationship to and understanding of space and place.

facing page
2-3 *The Haunting*
production images

The Haunting
MONT ROYAL PARK, MONTREAL
The Summit of Spirits
Level Three
HOUDINI
TOMASINA
CELL TOWER (SUMMIT)
TOMASINA
HOUDINI
THE ALLAN SISTERS
ARTHUR ELLIS
THE CROSS
RADIO AND TV TOWER
END
The Devil's Pitchfork
Level Two
THE KILLER
CYRIL
MIDGE
START
The Forest of Shadows
Level One
THE BELVEDERE
Olmstead Trail
Trail
Footpath
Bluetooth Beacon
GPS Area
Boundary Warning
Gateway

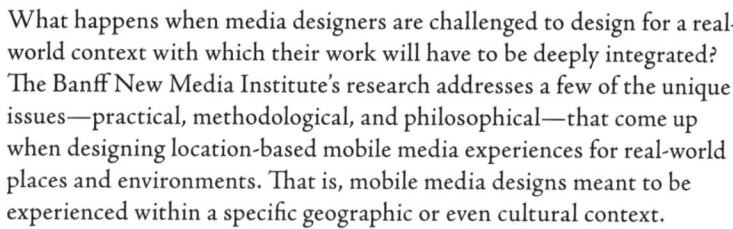

Tracklines
mobile media and the problem
of knowing the world

Angus Leech
Banff New Media Institute

What happens when media designers are challenged to design for a real-world context with which their work will have to be deeply integrated? The Banff New Media Institute's research addresses a few of the unique issues—practical, methodological, and philosophical—that come up when designing location-based mobile media experiences for real-world places and environments. That is, mobile media designs meant to be experienced within a specific geographic or even cultural context.

As mobile experience designers in Banff National Park, nearly every facet of what we build as part of the *Tracklines* project has to take the local environment into prior account and be tested in-context to find out how it behaves in combination with the world. Ultimately, our designs succeed or fail based on what happens when they are introduced into the landscape they were designed for: how they integrate with it, resist it, or interact with it—in effect, what kind of conversation ensues.

So what happens when we are challenged to understand the implicit language of unfamiliar environments in order to have this conversation? When locative media designs are compelled, like a form of sustainable architecture, to integrate with the existing patterns or dynamics of a place? What happens when people are no longer just interacting with media systems, but when people, media systems, and the surrounding environment all interact with each other simultaneously? Ultimately, what does it mean when locative media design practice also implies the practice of knowing a place, the practice of knowing a community, and the problem of knowing the world?

1 ***Tracklines*** mobile researchers on Banff's Hoodoos Trail

facing page
2 **An evolution** of screen graphics from the *Tracklines* mobile prototype

Everything is All Going on at the Same Time

Nina Wakeford
Goldsmiths College, University of London

Every weekday, hundreds of bicycle couriers circulate around London, delivering packages containing documents, media for film and television, and a plethora of objects which can move around the city in a courier bag. Some companies in large offices may call a courier to transport items between floors in the same building. Others may regularly keep couriers waiting, despite the charges mounting up, because they forget that these cyclists often arrive only moments after the initial call, having made their way through traffic at a speed impossible by any other means.

Thinking about the experience of cyclists—both couriers and commuters—can help us innovate both our theoretical and methodological frameworks for studying mobility. My title is taken from an interview with a courier who explains, 'We're calculating a thousand different things, whether it is the angle of the light, the surface of the road, the hatch covers, whether it is wet, raining, or dry, how much weight we've got on our back. Everything is all going on at the same time.' Another comments that 'You never arrive... you are always going to be riding.' Such experiences suggest that we should focus on the visible and invisible work of doing mobility, rather than concentrate on getting between locations. This has methodological implications.

The empirical data presented at the Mobile Nation conference was derived from a four-week project funded by Intel Research People and Practices Research Group. On the academic side, we were interested in exploring the ways in which the project could explore new visual ways of knowing about mobility, and how they might be shared with our industry funders. Researcher Britt Hatzius staged a range of experiments gathering visual data using still photography, digital video, and mapping. These were developed into an installation set up at the funder's labs. New visual methodologies function not only as a way of gathering data, but also as a way to perform translations between actors and disciplines with interests in mobility research.

facing page
1-2 **Stills** from courier mobility research experiments.

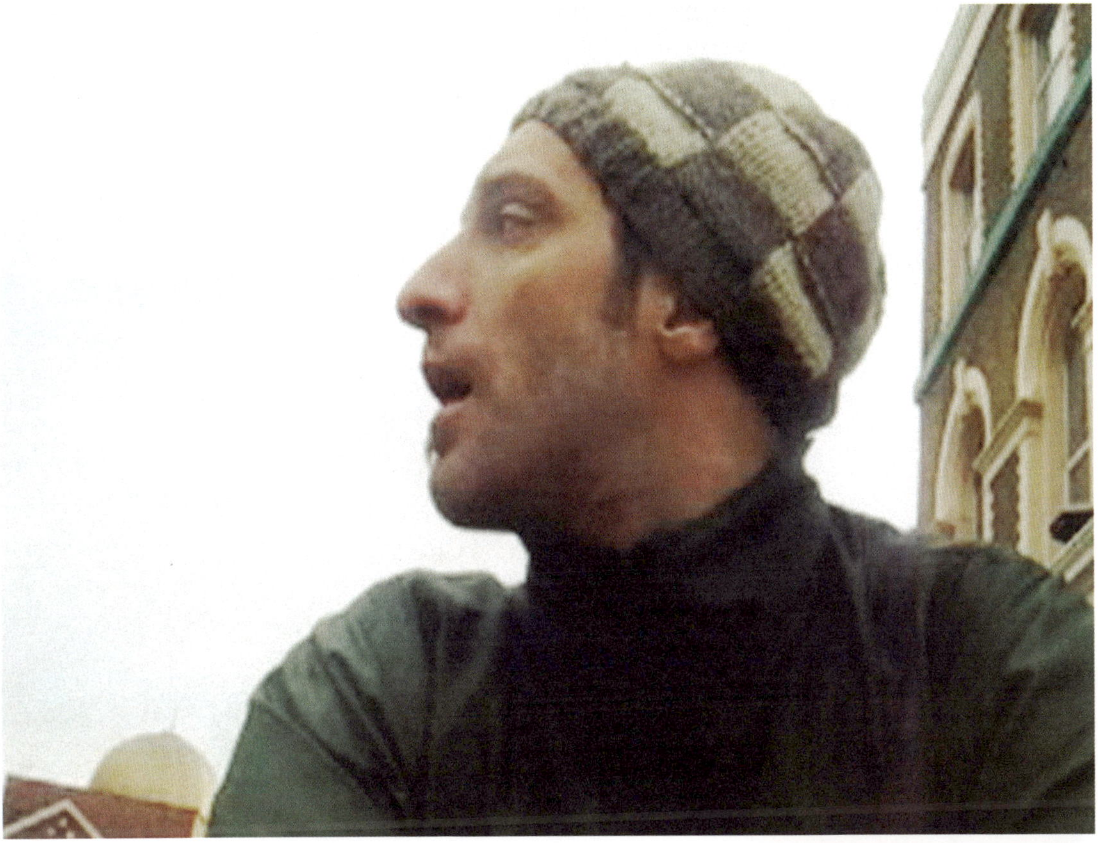

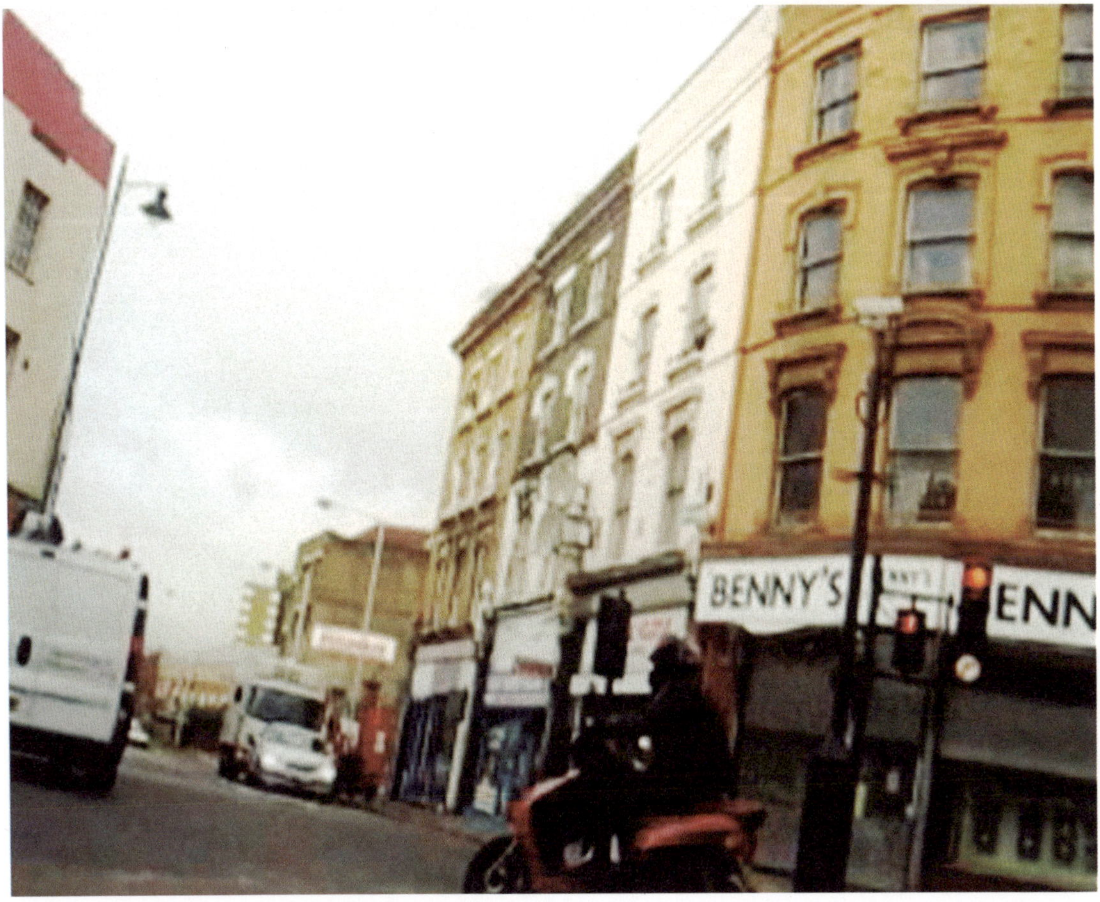
BENNY'S

Creating for the Multi-platform Context

Mobile content may consist of content ported from other disciplines and platforms, and increasingly we see adaptations of traditional media, broadcast media, and material from the World Wide Web for mobile devices. This content is key for industry aggregators and can be a valuable source of media for the mobile realm. Such content must, however, make sense within the social context of mobile phone use, and should be appropriately adapted for the mobile platform.

Questions addressed by designers, scholars, broadcasters, and industry experts are many: What types of methods are appropriate for designing location-based experience? How are mobile technologies designed and how are they adapted for multiple platforms? What are the changing objectives and uses within mobile communication design? In which contexts is mobility valuable, and when does mobile communication become intrusive? How do we design content to be both appropriate for multiple platforms and appropriately device-specific?

The migration of traditional forms of content to mobile platforms is challenging and requires a thorough understanding of perception, economies of scale, and context of reception.

Halos
making more room in the world
for new political forms

Nigel Thrift
University of Warwick

*Who would still dare to claim true ownership of his anger nowadays,
when so many take it upon themselves to tell him how he feels—indeed,
they know it better than he himself does?*
—R. Musil, *The Man Without Qualities*

He whose face gives no light, shall never become a star
—William Blake, *Proverbs of Hell*

Introduction

This paper is a part of a general attempt to struggle over the hill of various Western philosophies, social sciences, and forms of politics in order to see a new, more open vista, one in which, through the articulation of an ontology of achievement, different associations are able to be made and made manifest, different togethernesses are thereby able to be forged, and different landscapes of possibility are subsequently able to be uncovered.

I have fixed upon one particular visual device with which to try to make some sense of what I am trying to get at—the halo—which stands for a *change in the nature of the representation of imagination*,[1] on the grounds that a good part of the political consists of the establishment of an effective imaginary.[2] After all, it can be argued that, if power means the capacity to make somebody do what she would not otherwise do, then whoever possesses the capacity to influence others' imaginations has a good deal of power: in other words, political power is not only about controlling the means of coercion, but also about controlling the means of imagination, where imagination is understood as the ability to express possible/play/pretend beliefs and emotions that might become the basis of a better world and so-called imaginative resistance,[3] the possible beliefs and emotions that we resist imagining and accepting—either because we cannot imagine a possibility or because we do not want to imagine it.[4] These different dispositions and propensities might well be thought of as the basis of political-moral authority.[5]

The paper is therefore organized as follows: The introduction continues by sketching out the conceptual background which I will take as given in the discussion that follows. Subsequently, I will introduce the central conceit of the halo and then proceed to make an argument around three different uses of the halo: as a means of approaching affective imitation and its ramifications, as a means of understanding the generation of semiotic intensity and thereby affective traps, and as the construction of forms of community which attempt to generate affective affinity in new ways. In each case, my purposes are political. Respectively, they are to displace prevalent models of political activism, to understand new forms of inhabitation and their possibilities, and to generate new locatives.

As a prolegomenon, and as a general background to what I have to say, let me begin by stating what I think it is possible to argue we have learnt about the world of the social sciences and humanities over the last thirty years or so, understanding that our act of perceiving it is mixed up with our means of enacting it. I have identified nine interrelated assumptions that are sprinkled through the literature which, though hardly new, have now begun to produce a new and potent combination which is gradually becoming more than that.

First, that the world consists of associations that strive and swerve rather than simply follow causal laws. Thus, the future is never a static inversion of the present.[6] So, for example, life exhibits emergence based upon so-called *facilitated variation*, rather than selection.[7] After all, for there to be selection, there has to be variation. Such a viewpoint gives much more room to the organism but organisms are constructed to facilitate evolutionary change; that is, they are selected for *evolvability*: the organism participates in its own evolution. Thus, the capacity for facilitating variation has itself evolved.

Second, that much more of the world can be regarded as 'social'—assuming that we even retain the term—than we would once have thought likely or possible, not least because the presence of others does not just act as a source of direction but as a general energizer. To illustrate this point, let me take just one example: the humble cockroach. We now know that, for example, the speed of cockroach development is influenced by the quality of its social upbringing. We now know that the cockroach is motivated by the presence of its fellows to complete or carry out tasks more ably than it would if its fellows were not there: facilitated performances were produced when cockroaches were aware of audiences of their peers.[8] We now know that the mere presence of conspecifics is a source of general arousal.

Third, that societies are so much more complex than we have wanted to depict them as, to the point where it may be best to just drop the term 'social' in favour of compound terms like 'hybrids,' 'stuff,' or 'zoon.'[9] That for at least three reasons: One is that we know them to be linked to all kinds of other entities in conglomerates, condensations, or compounds using the 'natural' human capacity to fold which do not slavishly reflect social organization.[10] And this is before we get to the transgressive elements of embodiment of the Deleuzian kind. Another reason is because it has become increasingly clear that even small and apparently coherent societies can display all kinds of variation. Take the case of the Hopi:

> *The people called 'Hopi' are actually a loosely organized group of independent matrilineal clans. Each clan has its own set of rituals and ceremonies with each their own origin, efficacy, area of responsibility,*

and method of performance. Within each clan are individuals of varying age and status who can produce varying degrees of accurate statements concerning their religious ideas when interviewed. Not only are there disagreements between clan traditions, but also one finds conflicting traditions and institutions peculiar to Mesa and village affiliations.[11]

One other reason is because there is at least some evidence to suggest that social scientists have chronically underestimated the depth of the difference that is on display—a legacy of the Durkheimian emphasis on social facts, perhaps. That is true not only in the sense that there seem to be many hybrids and processes of which we are only dimly aware and for which we often do not have the requisite descriptions or 'landing strips' to recognize them. In other words, we are blinded by particular meta-phorical transit systems. Again, recent developments are laying out on display for us the sheer variety of passions that human beings can be motivated by. Recent developments on the Internet have allowed all man-ner of communities to coalesce which display every passion imaginable—and a good few that are not. Further, these passions are being played to through new systems of open innovation which both amplify and invest in them. In other words, the power of what might be called condensation has increased, allowing more kinds of collective intelligence to flourish but our vocabularies are still stuck with terms like 'crowds.' If I were looking for an analogy it might be with the discovery of plein-air painting in the nineteenth century and the host of new cultural styles it undoubt-edly made possible.

Fourth, we have to recognize that the exact nature of the results from social science and the humanities is continuously changing in ways which make it clear that it has no stable ontological or epistemological base. Rather, just as Western music has gradually accreted a repertoire of new sensings—chords, passing notes, dissonances, new forms of rhythm, bass, contrary motion, chromatic scales, inversion—so social sciences and humanities have been able to produce new awarenesses. Let me take just the example of language and how it frames our perception to make my point. It is still proving excruciatingly difficult to work that issue out. So, for example, tribes have been found which seem to have such simple lan-guages that they have a highly specific perception, one shorn of certain colours, notions of anything approaching number, and so on. Similarly, tribes have been found which have such intricately complex languages, based on Byzantine knowledge of genealogical structures and closure from other languages, that they seem to approach the limit of cultural transmission in a society without literacy or any significant division of labour.[12] Or take the by-now well-established turn to cognitive ethnogra-phy and anthropology and its leavings. In turn, the knowledge appropri-ated from these studies of distributed consciousness is up for grabs. Take the example of Hopi space and time. Since the 1970s, the original and apparently definitive writings of Whorf on linguistic relativism have been tested to destruction. For example, Whorf's work on Hopi space and

CREATING FOR THE MULTI-PLATFORM CONTEXT

time has been completely overturned. The work of Ekkehart Malotki, Armin Geertz and others has shown that Hopi space and time are nothing—and I mean nothing—like those originally depicted by Whorf.[13]

Fifth, we know that it is possible to invent new forms of encounter which work with and through the energizing effects of nonhuman others. Indeed, it would be possible to argue that the reservoir of matter that is, in some sense, 'alive' (a word that has itself has come to take on an increasingly blurred meaning) is increasing, assuming that it makes sense to talk of matter as live or non-live at all. Ever since Artaud's alchemical theatre, and no doubt before, the issue of designing new forms of encounter has been framed as an issue of shaping flux, a project which brings me to the issue of space.

Sixth, there is a move away from an obsession with exact localization. Spaces are all moving, in process, the result of what Sloterdijk calls the 'kinetic excess' of modernity: 'ontologically, modernity is a pure being-toward-movement,' a choreographic ontology. It follows that there is nothing fixed in the dominant 'choreopolitical' order, an order that requires extraordinary levels of resource to keep its 'most real' reality placed. So, for example, 'the process of subjectification in contemporaneity [can be seen] as that of an idiotic militarization of subjectivity associated to widespread kinetic performances of Tayloristic efficacy, efficiency, and effectiveness.'[14] Thus, the issue becomes one of thinking what a kinaesthetic politics would look like in an exhausting and exhausted onto-political project of 'being-toward-movement.' No wonder that some of the most incisive studies of space currently to be found come from within performance studies—a point I will return to later in the paper. At the same time, actual performance puts a check on some of the more exaggerated claims made by a modern generation of theorists who tend to provide accounts of the world that aim for a filling out of everything by showing how all manner of small but significant cavities exist in the dominant order, cavities that can be linked given the right conditions. There are many ways of going on, in other words. For example, as Pettit puts it, the range of possible transgressive perambulations that can be played with might include excursions, incursions, infections, expulsions, and fetchings, all of which remind us that space is never about exact location.

Seventh, the business of being-toward-movement is boosted by the extensive means of perceptive supplementation that now exist: a massive range of new means of extending sense-ings which are being slowly incorporated into practice by artists, by designers, by all kinds of everyday community. As a result, sense-ing is beginning to move away from being understood as simply a form of transmission of information about objects to a process which enables tangible qualities to be passed from one party or object to another, including the nonhuman, not least because the body itself is modified by all manner of technical interactions. Thus we begin to come close to a previous form of sense-ing, one in which 'the senses gave out information or affected others directly, as well as receiving information or serving as a conduit that might change an individual. In this way much that we might explain as miraculous or magical took place as part of the natural order.'

Eighth, and relatedly, given that, taken together, we are living in a time which is allowing a flowering of new kinds of social invention it therefore ought to be possible to find practical technologies and communities that can enable all kinds of twisted readings of the world which can challenge existing conduits of power by not just undermining but replacing them. To take up the spatial theme once more, this new kind of 'craftsmanship' will entail, above all, the forging of new locatives which allow different kinds of sensings to make their way into the world.

Finally, that somewhere in all this there is a reworking of ethics. Somewhat put off by claims that this will involve the rediscovery of piety, we might call this a general ethics based on a corporal compassion which transcends the human subject in that it forms obligations to more things than human, by entering into knots which can only be achieved in concert with the nonhuman. As will become clear, I am in favour of an approach which values presumptive generosity, including a general commitment to the challenge and creativity of 'life,' understood as a much-expanded realm of animacy. But this move does not require a relapse into either a macho philanthropy or a romantic sentimentality. Rather, it requires the foregrounding of an orientation to 'how' one is obliged and what betrayal would thus consist of, in other words an ethos of experimentation rather than representation.

Halos

How to make sense of this vast and gathering storm of new sensings of the world? I have three interrelated goals in mind. To begin with, I want to trace out some of the changes in the political contours of our time, but I will do so by understanding social process as a mass of material entanglements slowly changing within daily practices, often without intentionality: 'the mass seems to move with a life of its own. But the movement is built from the little micro-details of life.'[15] Specifically, I will be tracing out, perhaps in accelerated form, the further history of entanglement with material objects in which these objects come to have greater and greater purchase on our lives, not only as the ability to construct continuity and to initiate change, but also as small things too easily forgotten.[16] At the same time, I have a second goal in mind, and that is pointing to certain ways in which Western academic thinking is fracturing, with interesting consequences. It is fracturing because it is realigning the roles of concepts, percepts, and affects, reworking space and time, and taking on new partners (animals, materials). The result is a different take on the human, society, and imagination which is difficult to disentangle but can perhaps be made sense of as a generator of different kinds of 'radical' politics.[17] I have one other goal in mind, arising out of the previous two. I want to think about the possibilities of new creatures, understood as new compounds of life that act in unforeseen ways,[18] and the new spaces in which such life can flourish, spaces which provide new lures to feeling, new powers to force thinking and invention, new schemes of purposeful-ness—or not purposefulness—which can provide different means of moving us/them. In particular, I want to think about the kinds of model of affective agency that might be possible and, simultaneously (since they cannot be understood apart) how they might be fostered by speculative spaces which call to, provoke, and invoke these agents. In other words, I want to talk about new forms of intelligibility.

I will try to achieve these three goals by using the conceit of the halo— standing, in general, as a means of beginning the process of considering and constructing new imaginative *plausibilities*. More specifically, I will use the word 'halo' because it conjures up an image that has several kinds of grip. If I was to frame my argument in quasi-religious terms then what is being sought through the agency of the halo is a device which will not so much unite as bring into correspondence that which is different with-out trespassing on that difference and without trying to reduce what is puzzling to a predictable encounter. Rather, since 'each party may enter-tain its own version of the agreement,' the *art is in the achievement itself*. For my purposes the conceit of the halo opens up three specific dimen-sions of this act of *diplomacy* each of which is important for what I want to lay out in front of you, namely, the emergence and nurturing of infec-tious relationships, the design of space as both semiotic intensities and affective traps, and the construction of new kinds of community.[19]

The First Halo

To begin with, and most familiarly, of course, the halo is a staple of Christian religious iconography. Yet, the halo is pagan in origin. Many centuries before Christ, it is thought that various peoples of the Mediterranean decorated their heads with a crown of feathers: 'They did so to symbolize their relationship with the sun-god: their own "halo" of feathers representing the fan of beams splaying out from the shining divinity in the sky.[20] Indeed, people came to believe that by adopting such a "nimbus" men turned into a kind of sun themselves and into a divine being.' Various pharaohs and emperors followed suit. Later, the halo appeared in the art culture of ancient Greece and Rome,[21] before being incorporated into Christian art sometime during the fourth century, being given first to Christ, later to angels, and eventually to saints.[22] Subsequently, the halo has had a rich history as the aureole that appears to emanate from beings of particularly intense spirituality, a history with its own shifts in representation—for example, during the Renaissance, when rigorous perspective became essential, the halo changed from an aura surrounding the head to a tilting disc that appeared in perspective, floating above the heads of saints, and then to a thin ring of light.[23] In later work, halos would often appear by allusion or insinuation—as a cir-cular pattern that falls behind a head, or as an arc of a doorway.

In this paper, I want to understand the halo, first of all, as signifying the construction of new forms of empathy, that is both the act of identifiation with the feelings, thoughts, or attitudes of another and the imaginative ascription to an object, as a natural object or work of art, of feelings or attitudes considered to be present in oneself. In other words, the halo stands for two things simultaneously.

The reason is that I think that something quite interesting is happening in Western thinking of late. It is, I believe, a result of the joining of certain strands of thought as a result of more general changes taking place in the nature of the apprehension of space, thus pointing the way to a new kind of political settlement, one which might allow a different kind of spiritedness to emanate, one based on an ethos of craftsmanship of the moment that can produce 'instant' affective communities. In making this claim, I therefore want the halo to stand for an affective ambition which is the achievement of an infectious relationship.

To stake this particular claim, I want to fix on the halos to be found in the works of one the very finest orchestrators of glances, gazes, and stares, namely Giotto. I will start by examining one of Giotto's remarkable frescoes, The Meeting at the Golden Gate.[24] In this fresco, the aging Joachim and Anna, Mary's mother and father to be, look each other straight in the eye in an atmosphere of solidity and stability. The halo which unites them is an expression of this atmosphere of happy encounter. Nothing could be more different in another of Giotto's other finest frescoes, The Betrayal of Christ. There, only Jesus and Peter have halos. Judas does not: he remains a lonely and inner-directed subject, cut off from the affective flow. But I do not want to draw the obvious conclusions here about Euro-American Cartesian subjectivities, and the like. Rather, I want to fix on the face and how it is figured in these and other representations.

The uncanny semaphore of the face is a crucial element in both paintings and it will be a recurring motif in this paper. After all, 'the living face is the most important and mysterious surface we deal with... Babies just nine minutes old who have never seen a human countenance, prefer a face pattern to a blank or scrambled one.'[25] Almost from birth, the gaze is fixed on the face, especially the eyes, as the baby constructs joint attention and intentional understanding.[26] In other words, the face, like language, is an aspect of public thought.

It would be possible to make many different kinds of intervention concerning the face. But I want to fix on three. First, faces are one of the chief ways in which affect is generated in the world. Their forty-six separate muscles, their eyes, their mouths, their noses, and their ears allow a range of expression which is without peer in the natural world and they produce or certainly enable many of the characteristics which are most notably human: 'co-operation, commensality, morality, and the inhibitions that underlie it, prolonged dependence of offspring, capacity for intention attribution, planned deception, and the highly structured nature of social interaction.'[27] Second, there is the history of the representation of the face. Certain facial states come to be increasingly represented over the course of history. For example, the smile figures more and more as a result of the increasing portrayal of the open mouth from the eighteenth century onwards due to changes in social attitude—and better dentistry![28] Third, there is a technical history of the face, perhaps best illustrated by the history of cinema and its effects on our perception. For example, the close-up is a crucial way station in the history of the modern face, providing new means of attending to the face and new possibilities for relation, not least those arising out of the close-up's peculiar ability to generate both intimacy and threat, not least as a disembodied affect. The face itself becomes a frame. Thus, 'a face in close-up makes it possible for the spectator to generate hypotheses about the mind and feelings of the person depicted and ability to get psychologically intimate with him/her' but it can also be located outside the subject in the world of technically assembled images.[29] What seems evident is that the face is a crucial element of politics and the political. It was always thus, one might say. But the modern media have extended the range of body language in ways hitherto unforeseen, most especially by providing a set of stock affective scripts for which the face provides both the template and the chief means of operating.[30]

Most importantly, of course, the face is our chief means for producing and scripting affective effects. Through its medium, we exercise the capacity for mind reading that probably does most to distinguish us from animals.[31] Other creatures undoubtedly have pains, expectations, and emotions but having a mental state and representing another individual as having such a state is a second-order phenomenon which, so far as we can tell, other creatures do not have or have in an attenuated form.[32] Currently, the favoured explanation for mind reading is the so-called simulation explanation which effectively argues that

> people fix their targets' mental states by trying to replicate or emulate them. It says that mindreading includes a crucial role for putting oneself in others' shoes. It may even be part of the brain's design to generate mental states that match or resonate with, states of people one is observing. Thus mindreading is an extended form of empathy.[33]

In turn, the phenomenon of mind reading points to the importance of what I have called the infectious relationship as founded in the production of chains of imitation. This is a phenomenon that was noted early in the history of philosophy and psychology. For example, both Hume and Smith detected it in their writings on sympathy. Thus for Hume,

> When we see a stroke aimed, and just ready to fall upon the arm or leg of another person, we naturally shrink and draw back on our leg and on our own arm... The mob, when they are gazing at a dancer on a slack rope, naturally writhe and twist and balance their own bodies, as they see him do.[34]

Similarly, for Smith:

> When we have read a book or poem so often that we can no longer find any amusement in reading it by ourselves, we can still take pleasure by reading it to a companion. To him it has all the graces of novelty; we enter into the surprise and admiration which it naturally excites in him, but which it is no longer capable of exciting in us; we consider all the ideas which it presents rather in the light in which they appear to him, than in that which they appear to ourselves, and we are amused by sympathy with his amusements which thus enlivens our own.[35]

In particular, understanding affective relationships of this kind means understanding affective contagion, a central concern of turn-of-the-nineteenth-century social science in the form of the study of imitation and suggestibility. Imitation and suggestibility took shape as particular kinds of objects through a hypnotic paradigm which was worked out through an interest in particular forms of psychopathology (such as hallucinations and delusions), and an interest in spiritualist forms of communication. Imitation and suggestibility were sites for exploring all manner of issues, such as consciousness, memory, personality, and communication. In particular they signified a 'taking over' of the subject that defied normal economies of subject-object relations. However, subsequently, a move to psychoanalytic models of desire, or to more discursive approaches to subjectivity, ruled imitation and suggestibility out of court and they fell into disrepair as a way of approaching social structuring.

But, of late, imitation and suggestibility have been making a return, boosted especially by the rediscovery of the work of Gabriel Tarde on a somnambulist society and more general work on the construction of collective intelligence. Within cultural theory, viral models of contagion have been posited as explaining the workings of a range of phenomena, including ideology, governance, self-cultivation, and even resistance but often in highly speculative ways that posit a kind of performative energetics but without usually specifying what the source or content or form of that energy might consist of. But there is no need for this (often convenient) opacity, as I hope to show in this section through a more detailed examination of the grip of affective contagion.

Let me begin by summarizing what we know about affective contagion. To begin with, that means understanding affect as is in large part a biological phenomenon, involving embodiment in its many incarnations, but a phenomenon that is not easily captured via specular-theatrical theories of representation.[36] It brings together a mix of a hormonal flux, body language, shared rhythms, and other forms of entrainment to produce an encounter between the body (understood in a broad sense) and the particular event.

Then, affective contagion is generally semiconscious, something not that far from William Harvey's 'certain sense or form of touch,' sensation that is registered but not necessarily considered in that thin band of consciousness we now call cognition.[37]

Further on again, affective contagion is understood as a set of flows moving through the bodies of human and other beings, not least because bodies are not primarily centred repositories of knowledge—originators—but rather receivers and transmitters, ceaselessly moving messages of various kinds on; the human being is primarily 'a receiver and interpreter of feelings, affects, attentive energy.'[38]

In turn, this depiction points to one more important aspect of affective contagion, namely the importance of space, understood as a series of conditioning environments that both prime and 'cook' affect. Such environments depend upon pre-discursive ways of proceeding which both produce and allow changes in bodily state to occur. Changes in bodily state require understanding that essentially autonomic hormonal and muscular reactions are continually transferring between people (and things) in ways that are often difficult to track. At the same time, they challenge the idea that the body is a fixed component of humanity. It might be more accurate to liken humans to schools of fish briefly stabilized by particular spaces, temporary solidifications which pulse with particular affects, most especially as devices like books, screens, and the Internet act as new kinds of neural pathway, transmitting faces and stances as well as discourse, and providing myriad opportunities to forge new reflexes.[39] Thus, concentrating on infectious relationships requires a cartographic imagination in order to map out the movement between corporeal states of being which is simultaneously a change in connectivity. Only a very limited range of spatial models currently exists which can understand flows of imitation/suggestion, mainly familiar cartographic motifs from diffusion studies, certain very general metaphors that have arisen from the recent emphasis in social theory on mobility, a range of models of the staging of space that can be found in performance studies (which are usually excellent at showing how affect is conducted in intimate situations but often tail off when it comes to mediated contexts), a set of artistic experiments with sites of affective imitation that have often used the possibilities of modern electronic media, and various kinds of conversation maps.[40] However, it is also clear that certain technological advances, and especially those to do with mobile telephony and the web, are making it easier to visualize flows of imitation, not least because they are themselves prime conductors.

Now we can also add in what we currently know about imitation and suggestibility. For imitation has become a paramount concern of the contemporary cognitive sciences, and this work is worth exploring in a little more detail, since it contains many insights. In particular, imitation is now understood as a higher-level cognitive function,[41] mirroring both the means and ends of action, and highly dependent upon the empathy generated in an intersubjective information space that supports automatic identifications. For example, just as Hume and Smith might have predicted, hearing an expression of anger increases the activation of muscles used to express anger in others, especially those muscles to be found in the face. There is, in fact, only a delicate separation between one's own mental life and that of another, so that affective contagion is the norm, not an outlier. What differs between different cultures is rather what is regarded as the result of agency. Thus, for Western cultures it can be a painful realization to understand how little of our thinking and emotions can in any way be ascribed as 'ours'; it is very hard for Westerners to accept that broad imitative tendencies apply to themselves, both because they are unconscious and automatic, so that people are not aware of them, and because the preponderance of apparently 'external' influences threatens the prevailing model of an agent as being in conscious control of themselves.

At the same time, it is important to stress that imitation is more than mere emulation. Imitation is different from simple emulation in that it depends upon an enhanced capacity for anticipation, so-called mind reading.[42] In particular, much of human beings' capacity for mind reading (whether this be characterized as inference or simulation) develops over years of interaction between infants and their environments, and involves processing the other as 'like me,' and the consequent construction of high-level hypotheses like deception. That is, it involves a form of grasping which is innately physical and non-representational since our privileged access is to the world, not to our own minds.

Whatever the exact case might be, most imitation is clearly rapid, automatic, and unconscious and involves emotional contagion, in particular (down to and including such phenomena as moral responsiveness). In particular, people seem to be fundamentally motivated to bring their feelings into correspondence with others': people love to entrain. What seems clear, then, is that human beings have a default capacity to imitate, automatically and unconsciously, in ways that their deliberate pursuit of goals can override but not explain. In other words, most of the time they do not even know they are imitating. Yet, at the same time, this is not just motivational inertness. It involves, for example, mechanisms of inhibition, many of which are cultural.[43]

In turn, it is clear that imitation generates a spectrum of affective states and most especially empathy, not only because the self-other divide can be seen to be remarkably porous but also because across it constantly flow all kinds of emotional signals. But this is a kinetic empathy, of the kind often pointed to in dance, a kinaesthetic awareness/imitation which is both the means by which the body experiences itself kinaesthetically and also how it apprehends other bodies.[44]

Having considered the infectious relationship through the medium of the face, I also want to use Giotto's Christian iconography of Joachim and Anna's gentle gaze to reflect upon the possibility of forming new kinds of activists which are not the militant, even martial, activists we have too often lighted on: those who are 'self-confident and free of worry, capable of vigorous, wilful activity.'[45] In particular, I want to get away from the remains of the model of what Benasayag calls the 'sad activist,' always intent on configuring a centre from which to think radical practices,[46] a model which puts so many off—not just the committed but also the uncommitted, for whom it can appear that all activists are 'know-it-alls.'[47] In other words, I want to talk about how it might be possible to face up to the world by generating new models of the activist which are not like Walzer's constant hero strong of mind and will and which may well be more effective for that—certainly, they are closer to the kinds of open, infectious relationships I have been attempting to describe in their desire to generate affective affinities which are open-ended, emergent capacities to empower, rather than fixed programs which can be handed down.

Most particularly, I want to think about generating new moral-political stances (using this word to point towards the political and the spatial as aspects of each other) which express a different model of the political subject, stances which blur the boundary between mover and moved which is so crucial to prevailing models of active agency.[48] The feminist literature (especially on the literature on feminine nature), anthropologists, and others have been attempting this re-engineering of the subject over many years and it raises intriguing questions about many things we hold dear.

In particular, I want to talk about the virtues of passivity as a means of seeking reform. In the midst of current world events, this will no doubt sound discordant but passivity, or so it seems to me, points to a different way of doing things, one which dates from early modern times, and one which relies upon a very different model of agency and a very different rhetoric of passions which are dependent upon understanding subjects as transmitters and receivers of infectious relationships. So far as the model of agency is concerned it is crucial to understand that for new creatures

> agency admits of more positions than 'autonomous agent'... In addition to the autonomous agent undermined by recent discourses, an 'agent' can also refer to one who acts for another ... This deputized 'agent' is not a 'sovereign ruler' but a subject licensed by another authority to perform predetermined actions. The gap between 'agent' and 'autonomous agent' is crucial to seventeenth-century writers, who often deny 'autonomy' but insist on 'agency,' both descriptively (each individual has agency) and prescriptively (all individuals must act in the world). As 'agents' or 'instruments' of another, individuals are simultaneously 'acted by another,' in Thomas Hooker's phrase, and enabled to act in the world. 'Acted upon, we act,' summarizes John Cotton. These writers desire agency only insofar as it differs from autonomy: they desire not 'shaping power' over their identities and actions but to be shaped by another power.[49]

So far as the rhetoric of passions goes, what is important in becoming a new creature is the mobilization of passions like pride and humility politically, 'with the apparently "active" vice of pride condemned for its ineffectiveness and the "passive" virtue of humility serving the most dramatic revolutionary ends.'[50] The religious model of a radical that was prevalent in the early modern period was connected to the practice of a feminized humility: the agent was an instrument, 'the product of humiliation, anxiety, and soulful, feminine passivity, in the best sense of the word,'[51] an agent 'humiliated for collective sins past and reformed for the time to come': this is a 'feminine' passivity but not, I hasten to add, in any pejorative sense. It is fragility as a precondition of grace, passivity as a precondition of change.

What difference might this make? I am not sure. But take one classic example of a moral-political code in which the model of a constant, militant hero currently holds sway: courage and bravery.[52] Yet such a model varies markedly from what is considered seemly in other cultures. Our model of courage and bravery has its genesis in Greek notions of character. For example, for Aristotle, for every character trait, there is a vice of excess and a vice of deficiency:

> Aristotle says that true excellences of character—what are called the virtues—have in common that they strike the mean between excess and defect. Given a particular life-challenge, a courageous person will act in a way that avoids the excess of foolhardy recklessness, on the one hand, but also the defect of cowardliness, on the other. The courageous person will in any given circumstances be able to find an appropriate way to behave courageously. That is what it is to strike the mean: to find an appropriate way to behave in circumstances in which it is possible to do too much or too little.[53]

In other words, bravery is a virtue falling somewhere between rashness (bravery in excess) and cowardice (bravery in deficit).[54] But what constitutes bravery and what constitutes too much or too little of it varies massively across cultures and through time. Since for much of the time bravery and courage are clearly forms of 'thinking without words,' depending very much on taking a stance to a situation, they rely upon material symbols arranged as determinate spatial patterns for articulation. In other words, bravery and courage tend to be carried in particular material cultures which both illustrate and compound particular convictions: materials matter. I want to highlight two particular, contrasting examples of material cultures of bravery and courage, in which these virtues are thought through and as particular exceptional things/spaces, to make this point.[55]

For the warlike North American Crow people, to take one instance, bravery was materialized in the, to us, excessive practice of counting coup by planting coup sticks, that is, tapping an enemy with a coup-stick before killing him,[56] and then counting the coup in a ceremony after the incident in the form of a feather for each incident, which could be worn in the hair or on a shield. But this was a particular kind of bravery:

> Obviously, the practice of counting coups valorized bravery—a trait that was necessary for the Crow to survive. Honour was accorded to the brave men, along with access to women, extra food, and other material benefits. Imaginative-desiring-erotic-honour-seeking-life was organized around this kind of bravery. Little boys would play at counting coup, and little girls

would dance with the 'scalps' that their brave boyfriends had brought them. But if we look at Plenty Coup's list, we see that more was at stake than mere physical or even social survival. If the survival of the Crow tribe as a social unit had been the primary good, one might expect that the highest honour would go to the warrior who killed the first enemy in battle, or the warrior who killed the most. But to count coup it was crucial that, at least for the moment, one avoided killing the enemy. There is a certain symbolic excess in counting coups. One needed not only to destroy the enemy; it was crucial that the enemy recognize that he was about to be destroyed.[57]

In other words, this practice instantiated a particular kind of subject who could live up to the relevant standards of excellence associated with Crow culture.[58]

Take another instance. But here, the example is passive and arises out of a long tradition of non-violence, or, more accurately perhaps, the bravery of *hesitation*. I am thinking here, in particular, of standards of excellence lived by and instantiated in the Quakers.

> *If hesitation gathers practitioners, it is because rules and norms are discursive expressions tentatively formulating something that has no definitive, authoritative formulation and, hence, does not communicate with obedience—which I call 'obligations.' Obligations communicate with the possibility of their betrayal. If ever a practice exhibited this possibility, it is that of the Quakers who… did not quake in front of their God, but in front of the possibility of silencing what was asked of them in a particular situation, answering it in terms of pre-set beliefs and convictions.*[59]

Just like the Crow culture of bravery, so the Quaker culture of bravery is instituted by a material culture: the material interface is the meeting house which affirms the value of hesitation through the construction of an absolutely democratic space. In particular, the early North American meeting houses were built to a circular plan, thus producing an egalitarian acoustic in which everyone could be heard with the same volume wherever they spoke in the building.[60]

These are radically different, even opposed examples. But, in combination with the previous discussion of passivity, they lead to some intriguing questions. What should we count as bravery? Is there a political economy of bravery? Might it be possible to re-engineer bravery towards a 'passive' affective model again, so that many more exemplary acts could be included?[61] What kind of material culture would be able to achieve this? What does bravery look like in a conglomerate of relationships which includes all manner of material and animal correspondents? Whatever the answer might be, *space* will be key, and so it is to space that I now turn, and especially to the multifarious spaces being formed by various forms of contemporary information technology, and most especially to the spaces being formed by ubiquitous/ambient/pervasive/ persistent interfaces of various kinds, interfaces with which it is possible to have unconscious or semiconscious relationships.[62]

The Second Halo

Nowadays, the word 'halo' means as much to a Western audience as a bestselling series of computer games and associated comics and graphic novels,[63] with a fanatical—and I do mean fanatical—fan base. Based on the old science fiction conceit of humans versus aliens on a halo-like ring world, the *Halo* series first appeared in 2001, very much associated with the co-operation between Microsoft's XBox and the games developer Bungie Studios, which released *Halo 3* on September 25, 2007.[64] To give some indication of the popularity of this series, *Halo 2*, launched in 2004, has sold more than seven million copies worldwide so far.

Halo signifies the construction of world upon world. It is a series of terraforms, models of possible worlds. This seems extraordinarily important to me in that it presages the kind of world that is now coming into being, one based upon new disciplines like reflexive architecture, interaction design, environment art, and various forms of gaming which aim to redesign interaction. These disciplines allow passions that would have been difficult to express collectively to come into being through the design of new kinds of environments—synthetic worlds—that both facilitate play and close it down. I think it is no coincidence that there is currently so much attention being paid to new, more active forms of materiality: in a sense, these are the building projects of the twenty-first century since they presage a time when 'there really is no barrier to a complete translation of every human interpersonal phenomenon on Earth into the digital space' with all manner of results,[65] from new zones of economic activity through to new forums for interaction. And these are new realities. After all, as Castronova puts it, 'What happens in these worlds is not just play, and not just communication. It is a complex thing, a combination of real interaction and a play-like context.'[66]

Thus, in *Halo*, the purpose of the game is to move the characters through vast outdoor and indoor environments that have been imagined in great detail. Whilst the environments are designed by concept artists and executed by teams of designers who want to make these worlds 'look and feel real,'[67] they are also open to fan feedback. The environments are themselves characters in the game, what the designers call 'silent cartographers.' Objects are always also locations. What we see is the construction of new fields of occurrences and the construction of new entities that can count as events.

I want to suggest that *Halo* stands for a particular aspect of the modern world, namely a shift in the nature of mediation towards 'worlding' enabled by new material cultures which allow the affective priming of space to be systematized in ways which were not possible before. The game is symptomatic of the new 'stickiness' that is now possible in three ways.

Thus, on one level, it stands for how modern business has moved on from a focus on producing objects to a focus on producing worlds which must also inevitably be spaces. Thus, the business enterprise does not create its object but the world within which the object exists. As a corollary, the business enterprise does not create its subjects (as happened in the older disciplinary regimens) but the world within which the subject exists. Thus, as Lazzarato puts it,

> The company produces a world. In its logic, the service or the product, just as the consumer or the worker, must correspond to this world; and this world in its turn has to be inscribed in the souls and bodies of consumers and workers. This inscription takes place through techniques that are no longer exclusively disciplinary. Within contemporary capitalism the company does not exist outside the producers or consumers who express it. Its world, its objectivity, its reality, merges with the relationships enterprises, workers, and consumers have with each other. Thus the company, like God in the philosophy of Leibniz, seeks to construct a correspondence, an interlacing, a chiasm between the monad (consumer and worker) and the world (the company). The expression and effectuation of the world and the subjectivities included in there, that is, the creation and realization of the sensible (desires, beliefs, intelligence) precedes economic production. The economic war currently played out on a planetary scale is indeed an 'aesthetic war.'

The corporate aim is to produce and harvest what might be called decisive moments of affectively inspired semiosis which can be played into through the redesign of environments. Such engrossing moments have a deeply engrained cultural history, of course. The decisive moment was, in large part, an invention of Renaissance painters trying to depict major turning points in history. They would build up scenes in great detail in which the disposition of every person and object counted as a part of a moment straining towards realization. The motif was subsequently taken up by photographers, and especially photojournalists. Famously, for Henri Cartier-Bresson, the decisive moment (the title of his exhibit at the Louvre, the first photographic show ever to be so honoured) is the instant when a shutter click can suspend an everyday event within the eye and heart of the beholder producing a confluence of observer and observed. It is the 'simultaneous recognition, in a fraction of a second, of the significance of an event as well as the precise organization of forms which gives that event its proper expression.' Then, the decisive moment is still very much a mainstay of modern drama. Whole productions have been built around articulating the power of one moment, as in Deborah Warner's ability to focus the whole of *Titus Andronicus* on the moment where the raped Lavinia comes onstage, having had her hands chopped off and her tongue cut out. Her uncle, seeing this train wreck of a woman, seemingly incomprehensibly bids her good day and asks her where her husband is. The moment is often cut from the play by directors as impossibly discordant but Warner made it into a triumph, the key being, as she put it 'doing the right thing at the right time.' Finally, and most obviously, there is film. Cinema can be understood as a series of practical meditations on summoning up decisive moments: 'truth twenty-four times a second,' as Godard once put it. Cinema is able to produce not just speed but delay and deferral, preserving the moment at which the image is first registered in a kind of extended present.

On another closely related level, I argue that this game is symptomatic of the general rise of suggestible environments which can act to concentrate and guide infectious encounter by constructing traps for the affective flow of everyday life.[68] In turn, encounter can become a kind of currency, an insight that is drawn from the final writings of Althusser in which he refers to the genesis of a state of encounter, in which encounter is more and more able to be engineered so that it can be thought of as a kind of currency with a face value.[69] But perhaps, rather than drawing on a monetary metaphor, a metaphor of cultivation might be more appropriate. For there seems to me to be a direct line of descent between the knowledges of semiotic arrangement and disposition that landscape gardeners like Humphrey Repton thought to be so crucial to their art of making fictions manifest and the games of today.[70] These knowledges of arrangement and disposition are currently going through a new round of both strengthening and extension as evidenced by, for example, the general rise in cartographic awareness in all spheres of life and most especially by the experimentation with new forms of interrelation between mapping and the senses which are allowing infectious relationships to be both represented and engineered as never before.[71]

On a final level, these heavily corporatized suggestible environments signify a new sense of narrative which is not linearized.[72] A good example of this sense of narrative is provided by many modern games-influenced movies. Take *Pirates of the Caribbean: Dead Man's Chest*:

> *The film has no concern with cogent storytelling, and neither do today's youngsters. For them, fiction, like gaming, is an eternal present and plots a perpetuum mobile. The only narrative is to get to the next level. So while 'Pirates 2' spools for older people like a story whose reels have been muddled—a nightmare of botched narrative—for children and young adults, up to say, twenty, the film advances to higher things on stepping stones of incremental surrealism.*[73]

Overdone, no doubt. But, not that much overdone, perhaps. Derrida spent a considerable period of his career considering the way in which writing had imposed a particular form of linearization of time and space upon the world which was, in effect, the infolding of space and time known as 'book.' But Derrida was at pains to point out that linearization represents 'only a particular model, whatever might be its privilege' and he notes the increasingly obvious inadequacy of this model of arrangement to the 'delinearized temporality' and 'pluri-dimensionality' of contemporary thought; 'what is thought today cannot be written according to the line and the book, except by imitating the operation implicit in teaching modern mathematics with an abacus.'[74] The linearization provided by writing and its consequent inadequacy for certain kinds of thought can be thrown into relief by writing schemes that do not deploy the linear norm,[75] so-called non-discursive writings. There are many of these emblematic genres. For example, take the language of flowers, an early modern schema which used a variety of somatic registers—layout (e.g. circumference), colour, texture, smell—through which to display the special indistinction between natural objects and rhetorical figures.[76] This language made its way into many aspects of life as actual material objects, each of them understandable as utterances, from posy rings to nosegays, in a society which associated flowers with moral and other qualities. Viewed from our current perspective, such schemas as the language of flowers may appear to be inefficient codes, impoverished by a general lack of grammar and an unregulated 3-D, multi-sensory syntax which 'cannot be further combined into a restricted and therefore consequential utterance.'[77]

But, equally, from the perspective on linear narrative that is offered by some of the current developments, it is conceivable that a new form of narrative will be generated that is very close in form to the pre-modern prototype, one in which *conviction is carried in material objects and actions* (rather than what today is called the mind). Further, this is a sense of the world in which non-discursive writing is not readily distinguished from other human-made or naturally recurring patterns, wherein lies the recalcitrance to full referentiality which constitutes its particular force.[78]

This ambition is incarnated in Erasmus's celebrated description of a country house in which the walls, doors, galleries, flower beds, and wine cups are all decorated with improving messages,[79] an imaginative development intended to move on from the extant holders and transmitters of religious knowledge like the stained glass window to something all-encompassing.

> 'Who could be bored in this house', asks one guest, when among so many painted forms there is 'nothing inactive, nothing that is not saying or doing something?' Writing is positioned throughout the house and gardens to catch at the eye and activate the memory: religious texts and images remind the host and his guests of the way to salvation, and encourage them to pray; emblematic plants and animals carry various moral lessons; and painted birds and other trompe l'oeil effects cause wonder at 'the cleverness of nature… the inventiveness of the painter, [and] in each the goodness of God.'[80]

And it has never quite left the world. As Derrida put it, the linear norm 'was never able to impose itself absolutely,' not just because acts of cognition can occur outside it but because the linear norm is set to function as a limit and so opens the very questions it appears to close: the contingencies of graphic phoneticism, and the philosophical system that relies on it, depend upon an imposition that leaves in its wake all kinds of out of sequence gatherings that cannot be made to fit and that might be made to remind readers of the material practices that went into the production of the text. The ambition was kept in gardening, in some aspects of folk design, but there is more to this fugitive history. To illustrate this, I want to begin with Charleston House in Sussex,[81] the famous home of the Bloomsbury set, notable especially for Vanessa Bell and Duncan Grant's rich decorative style. Inspired by Italian fresco painting and the postimpressionists, the two artists decorated the walls, doors, furniture, and garden at Charleston to the extent that the house and garden became a living work of art in which every surface was semiotically enhanced, thus reproducing Erasmus's dream of a country house that would speak out from every corner. In doing so, Bell and Grant produced a mock-up of what the modern world is becoming like, a space in which even the marginalia are semiotically charged.[82]

But whereas their house and garden was an imaginative booster rocket which can be regarded as largely positive within its own bounds, much the same kind of ambition can also have profoundly negative consequences, as many totalitarian states have proved since. Such states exactly try to incarnate Charleston but as a space of propaganda. On this dark side what is crucial is to understand is the degree to which so much of the modern world consists of marginalia made central by so-called reactionary modernist forces. Thus Herf notes the way in which Nazi Germany attempted to design environments which would produce a total political experience by using media like radio, mass meetings, print media, and especially weekly wall newspapers which 'stared out at the German public for a week at a time in tens of thousands of places German pedestrians were likely to pass in the course of the day.'[83] Not coincidentally, about one third of the text of these wall newspapers seems to have been involved with anti-Jewish statements.

Of course, since the earlier part of the twentieth century, new visual technologies have run riot, technologies which both extend the means of representation (as in the proliferation of screens, the wall newspapers of the twenty-first century) and the registers which it is possible to decorate with images (as in the inhabitation of the precognitive domain by sigils like brands), thereby producing not just what Mitchell calls a 'networked vision,' but something closer to an electronic version of Erasmus's house in which every surface gives off continuously modulated messages, such that an exchange of qualities rather than just a transmission of information takes place—what Bruno suggestively calls a 'pandemic of images' that produces an 'aggregate mnemonic structure' that consists of multiple levels and planes of stimulation, disposition, and recollection, all jumbled together in various living reappropriations that constitute a kind of choreography, rising and falling to rhythms of its own.[84]

There is evidence to suggest that this process is gathering pace as a result of the intervention of large-scale parallel and distributed computation in all its forms which has allowed previously separate visual media—live-action cinematography, graphics, still photography, animation, 3-D computer animation, typography, and so on—to be *combined* in novel ways, producing what Rotman calls 'rampant visualism' and Manovich calls the 'velvet revolution.'[85] The underlying logic of this revolution which produces new media forms out of combination is one of remixability in which the computer simulates all media, thereby inducing a transformation of visual language towards 'motion graphics' that is 'designed non-narrative, non-figurative based visuals that change over time.'[86] What counts is the arrangement of elements like size, aspect, a line of type, an arbitrary geometric, another kind of form, and so on, into a kind of dance: 'we can compare the designer to a choreographer who creates a dance by 'animating' the bodies of dancers—specifying their entry and exit points, trajectories through space of the stage, and the movements of their bodies.'[87]

This is, as I have tried to make clear, much more than some putative society of the spectacle, that is an intensified deployment of the apparatus of the production of appearances in which

> the spectacle accelerates as a result of the falling rate of illusion; the disenchantment of the image-world may follow. In any case, we take spectacle in a minimal, matter-of-fact way to characterize this new stage of accumulation of capital. By no means just a piling up of images, as media studies would have it, but in Debord's sense of a social relationship between people that is mediated by representations. Crucially, our analysis depends on the complementary notion of the colonization of everyday life, and of subjection to an endless bombardment of brands, logos, slogans, consumption-motifs, invitations to feel happy. Globalization turned inward, as it were.

Rather than this state of fallen grace, what I am trying to describe is a reinhabitation, one based on making the environment—a word which itself becomes a contested one under the new conditions—into a semiotic soup but one in which most of the signs are non-discursive.[88] This reinhabitation is akin to the biosemioticians' notion that the basic unit of life is the sign.[89] It is not a direct imposition on a passive substrate of humanity but a reworking of what counts as through, a processual 'haptic spatiality.'[90] In other words, a non-representational mode of writing that utilizes non-linear deployments of time and space is again gaining a place in the world, with clear effects on what we regard as perception.

In turn, this change provides a pressing political task in a society in which this rampant visualism is coming into being. It poses obvious risks—but it also provides opportunities for building new kinds of community.

The Third Halo

In its third manifestation, the halo is a standard scientific term used to denote various optical phenomena that appear around light sources. It is therefore a natural term in sciences like meteorology, physics, and astronomy. For example, the galactic halo is a region of space surrounding spiral galaxies, including our galaxy, the Milky Way. It consists largely of old

stars, gas, and dark matter. It is believed that the galactic halo is a consequence of the hierarchical evolution of sub-galactic clumps seeded from
cold dark matter density fluctuations.

I want to use this image to return to the *matter* of imagination, understanding materiality as a series of occasions which are always moments of
knowledge. At the same time, I also want to think about the changing
shape of knowledge itself after the onset of information technology, as
knowledge increasingly takes on a significant non-paradigmatic halo
which cannot be centred or made a part of the whole[91] the result of not
just the expansion of knowledge but its increasing ownership by communities that have little or no relation to formal knowledge structures. In
other words, in this paper the third manifestation of the halo is as a
whole series of knowledges thought to be of little or no consequence
which form clumps of various kinds, a background which turns out to be
fundamental in seeding the universe of knowledge rather than incidental.
Why? Because I am convinced that these petty knowledges are a resource
that can be tapped to form a new political genre or genres,[92] one which
calls to and relies on affective contagion and which might be used to re-
engineer affective qualities like bravery and courage in productive ways.
This politics consists of clumps of like minds arranged loosely and indis-
tinctly in semi-directed practices which move beyond understanding
affective contagion as simple contact towards understanding affective
contagion as a kind of fluency practiced by design.[93] Let me make it clear:
this is not to suggest that if these practices were aggregated they would
suddenly form a new political force, but rather that they can form a different kind of choreographic strain, a contrary motion which both works
with and against the grain of 'being-toward-movement' and which might
allow us to sense and even construct new affective strains.

Set against those who think that 'our stunted imaginations have
largely lost the ability to think what a society other than capitalism...
might look like,'[94] I think we are living in a time of extraordinary imaginary outbursts if only we had the nous to touch and feel them, imaginary
outbursts founded in the co-operative symbiosis provoked by new situations, imaginary outbursts that force thinking by producing *affective affinities*. These outbursts have already had considerable purchase on the
world of mass daily practices but, on the whole, we are not picking them
up because they are based on 'discontinuities of pattern, the tiny causali-
ties of chance, the reparative and tender (as opposed to deadly and terri-
fying) features of intricate connection.'[95] They do not fit the standard categories—active/passive, micro/macro, passion/calculation, interest/
disinterest, objective/belief—we use to describe the world.

These outbursts could be named in all kinds of ways, no doubt. But I
want to draw on modern performance studies to try to describe them in
more detail. Performance has always understood the power of affective
contagion and sometimes has highlighted it. Think only of Artaud's
alchemical theatre:

> *What modern social science tried to make intelligible, Artaud tried to
> make real: the contagion of gesture, the communicative power of a scream,*

CREATING FOR THE MULTI-PLATFORM CONTEXT

a mimetic theatre of collective seizure and frenzied emotion, Artaud's intent was not to start a panic but experiment through performance with features of the social—never far from the alchemy of the theatre—that collective terror also opens toward. 'The mind's capacity for suggestion' which Artaud identifies as one source of theatre's transformative power, is precisely the capacity that modern social science locates as one source of the social itself.[96]

I will call these outbursts 'dances that describe themselves'[97] in order to give me a means of naming them and as a place to start from, as a piece crafted spontaneously in performance—in the moment. The phrase comes from the work of that well-known dancer-choreographer of improvisation, Richard Bull, and his allegiance to thinking on your feet, by choreographing while you dance, thus producing a leaderless community. But his dance was not just a piece of random improvisation, worshipping 'liveness.' Far from it. It was an act of possession—and command. The premise was that a set of dancers would come together and over several weeks would slowly tune their world view to the presence of 'the dance that describes itself.' This tuning was intended to unite their bodies (and, to an extent, those of the audience) in the flexing, undulating mass of 'the dance.'[98]

> *They moved in and out of 'possession,' enjoying the shift in perspective, the different sense of agency that becoming inhabited by The Dance allowed. The Dance told them what to do and they necessarily complied, yet they also created The Dance, determining when and how it might take control of their dancing. The contradiction between these two selves, one possessed and one in command, opened an ironic tension that reverberated throughout the entire performance, a tension compounded by the fact that the dancers described, often with clinical precision, their actions.*

> *Typically, the act of possession entails a loss of speech and the inability to describe during or afterwards what happened while dancing. The Dance That Describes Itself plays upon this venerable and ancient trope of giving oneself to the dance, becoming one with the dance, or being free in the dance. Yet it constructs a different kind of possession. Dancers are asked to remain highly conscious of their circumstances and to describe their actions verbally. Rather than serve as mute embodiment for cosmic forces, indescribable in their proportion and power, these dancers comment adroitly on mundane motives, frustrations, or desires. The collision between two incommensurate images of dance—one speechless and transcendent, the other analytic and pedestrian—reinforces the irony inherent in the initial proposition of being possessed.*[99]

I would argue that the practices that I want to describe are akin to this stance in that they involve a careful tending of knowingness through the design of empowering situations which are based on producing new and speculative *locatives*, indications of place *and* direction *and* affinity, which privilege an openness of form which is still, however, able to shape and mould and comment upon that process. At the same time, these locatives produce new time frames, new notions of 'calendarity.'

What would these new locatives look like? The history of performance undoubtedly gives us some clues, where performance is understood as the construction of socially and technically informed living entities, since in many ways it has been born out of an impulse to remap spaces and, in particular, to escape the constraints of enclosed theatrical spaces and the kind of conventions they abide by: discrete physical locations exploiting particular kinds of sound and lighting, linear manipulation of timelines through devices like reminiscence and premonition. Indeed Norman argues that the vestigial geometries of these spaces and times still hamper our ability to craft other kinds of social encounter. We keep on beating the same bounds, often unconsciously.[100]

Perhaps the most important step has been to get away from an obsession with exact localization. In the early nineteenth century, Victor Hugo had already pointed out the benefits of the strategy of localization in inducing a sense of reality, but also its risks in dictating imaginative content. The problem, of course, is that too often the reaction to the risks associated with localization has been to fall back on a notion of spontaneous gatherings of individuals, along with the common graphic vernaculars for depicting these instant multitudes—the crowds of which various forms of flash mob are often considered to be the latest manifestation.[101] The generation of apparently primal spontaneity has had a long history in performance, dating from at least Wagner's 'free associations of the future,' and it has had obvious political downsides. However, at the same time, it has also led to a very large amount of thought about how performance works at the pre-individual level and how the performer acts as an enhanced transmitter of various forms of sympathy, culminating in many acts of ecological theatre which try to conjure up a sentient unconscious, if that is not a contradiction in terms, through creative engagement of the feelings of the audience in the exploration of space.[102]

The tension between these strategies of localization and spontaneity and the knowledges that they produce is currently being worked out in 'post-dramatic' artistic performances that pull all sorts of beings into a communion of direct living perception that develops with and within time.[103] In particular, these performances explore the dynamics of affective emergence by constructing 'living organizations' out of the new locative media. That task involves maximum experimentation across many registers of the senses in order to 'feel' all the data available in a particular universe that might belong to an emerging entity,[104] using the full range of modern locative technologies as vital intermediaries. In turn, this task has generated what is often a calculated indifference to where performance is meant to take place. For example, performance can even be located in outer space, in the domain of so-called metagestural proxemics, as in the spacesuit that will be crammed with communication electronics and thrown out of the international space station to burn up in the atmosphere, or tometaxy.net's attempt to produce a collective public sculpture to world peace in orbit around the Earth and ultimately a moon installation, or Nam June Paik's moon. In other words,

> *locative media performances encompass participants and forge identities ranging from the most intimate to the most distant; the propensity of such performances to go global is equalled by their aptitude to inject highly localized, often time-bound events into overall connected fabrics. It is this tension between localized input on the one hand and web-borne, purportedly universal resonance on the other that gives mobile systems their complex social and artistic potential.*[105]

In practice, this means that post-dramatic theatre attempts to produce performances across many sites simultaneously in what we might call a choreopolitics which is backed up by a range of different practices which blur the divides between what used to be known as art, as politics, as social science method, and as information technology in a concerted attempt to batter down particular imaginative resistances through a mixture of 'displacement, dislocation, distribution, and disorientation.'[106] This blurring first occurred in the 1960s but, after a forty-year history, it now provides a body of formal and informal knowledge of considerable sophistication, ranging from the kind of project that simply utilizes user-led functional cartography (for example, Michelle Teran's *Life: A User's Manual* which tries to technologize Perec), through Jonah Brucker-Cohen and Katherine Moriwaki's *Umbrella.net*, and on to all manner of projects that chart out movement (such as Teri Rueb's *Choreography of Everyday Movement* where trails of dancers moving through the city are tracked with GPS to obtain real-time dynamic drawings which are then printed on to acetate which is sandwiched between stacked glass plates that grow taller and more complex with each addition). One might argue that such projects act as nothing more than twinkling marginalia. That would, I think, be a mistaken judgment. Rather, I think they are attempts to call up the outlines of spaces which can produce affective shifts which, though they may appear to be minor, can have major effects, for example spaces that administer a shock to irritation so that it moves from a space in which over-annoyance is accompanied by insufficient anger to something more affectively productive—such as the bravery required to actually intervene.[107]

To end, I will fix on just two examples of this new form of mobile acting up, examples which tap into a long tradition of artists acting as agents-provocateurs investing wider communities,[108] examples which aim to produce politically charged works of art that exist outside any extant paradigm as a kind of kinetic outcropping or even stray which exhibits no deeper truth than the achievement of an affective phase-shift.[109] But that, of course, is the point.

The first example is the *Displaced Emperors* project. *Displaced Emperors* was the second relational architecture project. This installation used an 'architact' interface to transform the Habsburg Castle in Linz, Austria. Wireless 3-D sensors calculated where participants pointed to on the facade and a large animated projection of a hand was shown at that location. As people on the street 'caressed' the building, they could reveal the interiors, which corresponded to Chapultepec Castle, the Habsburg residence in Mexico City. In addition, for ten schillings, people could press the 'Montezuma button' and trigger a temporary post-colonial override consisting of a huge image of the Aztec headdress that is kept at the ethnological museum in Vienna.

The second example is *Ballettikka Internettikka*. *Ballettikka Internettikka* is a series of tactical art projects which began in 2001 with the exploration of Internet ballet. It explores wireless Internet ballet performances combined with guerrilla tactics and mobile live Internet broadcasting strategies. Its Internet guerrilla performance has mainly consisted of invasions of particular art-houses, such as the Bolshoi Theatre in Moscow (March 2002), La Scala in Milan (November 2004), and the National Theatre in Belgrade (October 2005) and the mounting of alternative Internet performances within their confines, on the premise that the passions of opera can be transferred to the 'passionless' Internet, so producing radical emotions. In 2006, the two main artists, Igor Stromajer and Brane Zorman, performed a new guerrilla net ballet, this time in the men's toilet in the basement of Volksbühne Berlin, using a flying cow, a group of small robots—and a toilet seat. The artists utilized low-tech mobile and wireless equipment for the invasion and live broadcast: a public, unprotected wireless Internet connection point, available for free at the Rosa-Luxemburg Square in Berlin, and free RealProducer and Live LE software for streaming video and audio with live manipulation.[110]

In one sense, these interventions are simply the latest chapter in the long history of trying to produce grammars of movement of the kind to be found in the history of dance since at least the sixteenth century. But, at another level, they are embryonic political interventions, affective utterances which, through the production of spatial and temporal coherences which are also new forms of imaginative assay,[111] are intended to boost encounter and thereby provide new means of animatedness and attentiveness. In particular, they are trying to struggle out from a notion of a place which is bounded—an 'environment'—towards notions of place as relationships with space that are rather like that of the face in their ability to be expressive and to reveal what the other is thinking: space as 'the eyes of the skin,' so to speak.[112] Thus space becomes richly emblazoned with signs of thought. In a sense, space becomes face.

Acknowledgements

I would like to thank Søren Buhl, Armin Geertz, Susan Hurley, Britta Timm Knudsen, Sally Jane Norman, and Minna Tarka for their comments on this paper.

Notes

[1] I take representation here to be able to be understood as doings rather than a relation between an inner and an outer. Our basic grip on the world consists not of inside out but of representing deeds: deeds are themselves representational. M. Rowlands, *Body Language: Representation in Action* (Cambridge: MIT Press, 2006).

[2] C. Castoriadis, *The Imaginary Institution of Society* (Cambridge: Polity Press, 1997).

[3] Imagination actually fires in the same area of the brain as belief and can generate equally strong affective reactions. Thus, most philosophers now count it as a 'distinct cognitive attitude,' different from beliefs in some respects but not others.

[4] Imaginative resistance can range from simple presuppositions, such that a nursery school room can be called a Kangaroo Room or a Bumblebee Room but not a Dung Beetle Room or a Vulture Room, through more serious cases such as resisting stories in which female infanticide is counted as morally right to the resistance of the German population in the Second World War to thinking about the annihilation of the Jews. See J. Herf, *The Jewish Enemy: Nazi Propaganda During World War II and the Holocaust* (Cambridge: Harvard University Press, 2006); S. Nichols, ed., *The Architecture of the Imagination: New Essays on Pretence, Possibility, and Fiction* (Oxford: Oxford, 2006).

[5] Of course, the imagination has routinely been figured as something that cannot be conventionally controlled but there are many reasons to think that this is only partially the case and that it can indeed be engineered, especially in an age of manufactured vision in which 'we now know ourselves in our mind's eye mostly by projecting a camera's eye view.'

[6] T. Lorraine, 'Feminism and Poststructuralism: A Poststructuralist Approach' in *The Blackwell Guide to Feminist Philosophy,* eds. L.M. Alcoff and E.F. Kittay (Oxford: Blackwell, 2006), 266–282.

[7] M.W. Kirschner and J.C. Gerhart, *The Plausibility of Life: Resolving Darwin's Dilemma* (New Haven: Yale University Press, 2005).

[8] R.B. Zajonc, A. Heingartner, and E.M. Herman, 'Social Enhancement and Impairment of Performance in the Cockroach,' *Journal of Personality and Social Psychology* 13 (1969): 83–92.

[9] C.A. Jones, ed., *Sensorium: Embodied Experience, Technology, and Contemporary Art* (Cambridge: MIT Press, 2006).

[10] Jones, *Sensorium.*

[11] A. Geertz, 'A Reed Pierced the Sky: Hopi Indian Cosmography on Third Mesa, Arizona,' *Numen* 31 (1985): 217.

[12] N.J. Enfield and S.C. Levinson, eds., *Roots of Human Sociality: Culture, Cognition and Interaction* (Oxford: Berg, 2006).

[13] E. Malotki, *Hopi Raum* (Tubingen: Narr, 1979); Geertz, 'Ethnohermeneutics and Worldview.'

[14] A. Lepecki, *Exhausting Dance: Performance and the Politics of Movement* (New York: Routledge, 2006), 13.

[15] I. Hodder, *Catalhöyük: The Leopard's Tale*. (London: Thames and Hudson, 2006).

[16] J. Deetz, *In Small Things Forgotten*, (New York: Anchor, 1996).

[17] L. Tønder and L. Thomassen, eds., *Radical Democracy: Politics Between Abundance and Lack* (Manchester: Manchester University Press, 2005).

[18] I prefer the word 'compound' to the word 'hybrid' because it implies familiar concepts fused rather than genuine concept creation.

[19] Stengers is intent on describing the practice of shuttling between parties that disagree: 'The art of diplomacy does not refer to goodwill, togetherness, the sharing of a common language or an intersubjective understanding. Neither is it a matter of negotiation between flexible humans who should be ready to adapt as the situation changes. It is an art of artificial arrangements that do not exhibit a deeper truth than their very achievement—the vent of an articulation between protagonists constrained by diverging attachments and obligations in situations where contradiction seems to rule, a rhizomatic event without a ground to justify it, or an ideal from which to deduce it.'; regarding affective traps see Gell.

[20] S. Fisher, *The Square Halo and Other Mysteries of Western Art* (New York: Abrams, 1995). The need to preserve art objects may have also added to the development of the halo. 'Statues were kept not in museums but in the open. Therefore they were subject to deterioration through various causes. To protect them from the droppings of birds, the rain and the snow, a circular plate—either of wood or brass—was sometimes fixed upon their heads.'

[21] As the Roman emperors who began to think of themselves as divine beings, they wore a crown in public to imitate the sphere of light from the sun.

[22] Here, the halo could have a very detailed iconography. Thus, round halos were used to signify saints. A cross within a halo was used to signify Jesus. Triangular halos were used for representations of the Trinity. Square halos were used to depict unusually saintly living persons, still bound to earth and so not able to obtain the perfection of the circle. Allegorical figures such as the virtues wear hexagonal halos. And so on. For more see Fisher, *The Square Halo*.

[23] But it is important to point out that the halo is not confined to Western religious history. For example, it has been widely used in various forms of Buddhist iconography since at least the first century AD to depict the Buddha or Buddhist saints, a direct importation from the West to the East.

[24] The example is taken from Sloterdijk.

[25] D. McNeill, *The Face: A Guided Tour* (London: Hamish Hamilton, 1998), 4.

[26] N. Eilan, C. Hoerl, T. McCormack, and J. Roessler, eds., *Joint Attention: Communication and Other Minds* (Oxford: Oxford, 2005).

[27] Enfield and Levinson, *Roots of Human Sociality*.

[28] Jones, *Sensorium*.

[29] N. Hansen, *New Philosophy for New Media* (Cambridge: MIT Press, 2004).

[30] As this instance shows, the idea that a politics of radical difference has to entail a choice between networks of signification or networks of embodied matter seems overdone. Rather, recent work argues that embodied matter always has sign content. See W. Wheeler, *The Whole Creature: Complexity, Biosemiotics and the Evolution of Culture* (London: Lawrence and Wishart, 2006).

[31] It is now generally accepted that the brain developed in response to and as a function of social interaction and especially the ascription of intention—that

is, the attribution of actions, motives, intentions, and beliefs to fellow interactants—and that what we perceive is set up by the wiring of interaction produced by the set of most notably human abilities that plausibly evolved together, all of which were boosted by the enhanced communicative interaction arising from the paraphernalia of language—parsing, turn-taking, repair, and the like. The brain, in other words, has become an instrument of shared activity—an interaction engine rather than an individual setting. And, within broad parameters, this shared activity is remarkably heterogeneous, aided by the fact that the brain is in any case plastic so that particular experiences of shared activity act in particular ways, by the fact that systems of shared activity generate emergent properties, and by the fact that cultural variation is therefore more than just incidental but central to interaction. Enfield and Levinson, *Roots of Human Sociality.*

[32] Although, periodically, an animal will be found that has some of these features. For example, most recently, Dally et al. found that western scrubjays can keep track of which birds were watching them and what they might be thinking. If another scrubjay sees them hiding food, they move this stash when that particular bird is not present. They can distinguish individuals and imply a motivation: the desire to steal their hoard. See J.M. Dally et al. 'Food-caching Western Scrubjays Keep Track of Who was Watching Them and When,' *Science* 312 no. 5780 (2006): 1662–1665; S. Hurley and M. Nudds, eds., *Rational Animals?* (Oxford: Oxford, 2006).

[33] A.I. Goldman, *Simulating Minds: The Philosophy, Psychology, and Neuroscience of Mindreading* (Oxford: Oxford, 2006), 4.

[34] D. Hume, *A Treatise of Human Nature* (1739; London: Penguin, 1985).

[35] A. Smith, *The Theory of Moral Sentiments* (1759; Oxford: Oxford University Press, 1976).

[36] It is important to note that in this paper I am taking embodiment to be a linked, hybrid field of flesh and accompanying objects, rather than a series of individual bodies, intersubjectively linked. I take the presence of objects to be particularly important because they provide new means of linkage—new folds, if you like. See S. Zielinski, *Deep Time of the Media: Toward an Archaeology of Hearing and Seeing By Technical Means* (Cambridge: MIT Press, 2006); T. Brennan, *The Transmission of Affect* (Ithaca: Cornell University Press, 2004).

[37] In turn, it is worth remembering that the dynamic range of sensory nerves is startlingly poor: for example, they usually fire at no more than about 200 impulses per second, compared with, say, the fifteen log units of variation of intensity of light that the eye can deal with. So consciousness of whatever kind always comes heavily pre-treated (although all manner of tricks of information compression help to overcome some of the limited channel capacity of the sensory nerves to the brain).

[38] Brennan, *Transmission of Affect,* 87.

[39] For example, it is possible to write about the history of facial expressions like the smile because media have been invented which can transmit these expressions.

[40] J. Abrams and P. Hall, eds., *Else/where: New Cartographies of Networks and Territories* (Minneapolis, University of Minnesota Press, 2006).

[41] Thus, imitation has proved to be the rarer and cognitively more demanding ability in animals than trial and error.

[42] There is, of course, a lively debate in the cognitive sciences and primatology about what exactly is meant by mind reading (so, for example, some would have it that it requires the construction of full-blown beliefs about others' cognitive states, for example, something I think unlikely). And equally there is debate about

how far it stretches (so, for example, some apparent mind reading might consist of sophisticated behaviour programs). But, as Sterelny puts it, though imitation may not always be a 'theory of mind task, it is a cognitively sophisticated one.'

[43] It may even be, following Tarde, that memory and habit are forms of imitation: 'Engaged in either, we in fact imitate ourselves, instead of another person: memory recalls a mental image, much as habit repeats an action.'

[44] S.L. Foster, *Dances that Describe Themselves: The Improvised Choreography of Richard Bull* (Middletown: Wesleyan University Press, 2002).

[45] M. Walzer, *Spheres of Justice* (Oxford: Blackwell, 1988).

[46] Colectivo Situationes, 'Something More on Research Militancy: Footnotes on Procedures and (In) Decisions,' *Ephemera* 5 (2005): 602–614.

[47] N. Eliasoph, *Avoiding Politics. How Americans Produce Apathy in Every day Life* (Cambridge: Cambridge University Press, 1998).

[48] This distinction (mover and moved) can be traced back to Aristotle.

[49] S.P. Gordon, *The Power of the Passive Self in English Literature*, 1640–1770 (Cambridge: Cambridge University Press, 2002), 23.

[50] D. Gross, *The Secret History of Emotion: From Aristotle's Rhetoric to Modern Brain Science* (Chicago: University of Chicago Press, 2006), 110.

[51] Gross, *The Secret History of Emotion*, 93.

[52] Any web search for courage or bravery immediately produces vast numbers of military examples, showing the centrality of this conception of courage and bravery to our judgments.

[53] J. Lear, *Radical Hope: Ethics in the Face of Cultural Devastation* (Cambridge: Harvard University Press, 2006), 17.

[54] Aristotle does not mean that bravery is simply an average. Rather, the golden mean is different for each person, depending on their character and situation.

[55] A. Henare, M. Holbraad, and S. Wastell, eds., *Thinking Through Things: Theorising Artefacts Ethnographically* (London: Routledge, 2007). I could no doubt have chosen other examples—the warlike honour code of the Pushtun comes to mind as does the history of Gandhian non-violence, but these two examples seem to me to be striking enough to make the point.

[56] Counting coup could also mean taking an enemy's weapons while he was still alive, striking the first enemy to fall in battle, no matter who killed him, stealing a horse tied up in an enemy's camp, and so on.

[57] Lear, *Radical Hope*, 15–16.

[58] Crow culture was, for all intents and purposes, annihilated when the US government not only killed many of the tribe but also crushed the space in which Crow meanings might survive unchallenged, the very intelligibility of the terms in which people understood things to be happening—for example, by equating making away with an enemy's horse with horse stealing.

[59] Stengers, 'Including Humans,' 11.

[60] D. Rath, *How Early America Sounded* (Ithaca: Cornell University Press, 2003).

[61] I am thinking here of movements like the International Solidarity Movement which produced 'passive' heroes and heroines like Tom Hurndall and Rachel Corrie. 'Their acts of solidarity articulate a practical riposte to the despairing twentieth-century voices that wanted to discredit this sort of gesture by arguing that the openness and undifferentiated love from which it derives is

tainted, ignoble, and unpolitical.' It also points to another halo: the Halo Landmines Trust. See P. Gilroy, *After Empire: Melancholia or Convivial Culture?* (London: Routledge, 2004).

[62] T.W. Bickmore and R.W. Picard, 'Establishing and Maintaining Long-term Human-Computer Relationships,' *ACM Transactions on Computer-Human Interaction* 12 (2005): 293–327.

[63] Halo is also the name of a fictional superheroine published by DC Comics in the eighties and nineties. A film is also planned for 2008.

[64] There are also PC and Macintosh versions.

[65] E.W. Castronova, *Synthetic Worlds: The Business and Culture of Online Games* (Chicago: University of Chicago Press, 2006), 48.

[66] Castronova, *Synthetic Worlds*, 69.

[67] E.S. Trautmann, *The Art of Halo: Creating a Virtual World* (New York: Ballantine, 2004), 71.

[68] To some degree, large cities have always acted in this way, allowing 'epidemics' of imitation to be marshalled and directed, but what I am suggesting here is something with a much higher element of design.

[69] L. Althusser, *Philosophy of the Encounter: Later Writings, 1978–1987* (London: Verso, 2006).

[70] If Repton had been alive today, I suspect he would have been hard at work designing games.

[71] Examples of this new interrelation between mapping and the senses can be found in Jones's *Sensorium* and others.

[72] J. Fleming, *Graffiti and the Writing Arts of Early Modern England* (London: Reaktion, 2001).

[73] N. Andrews, 'The guess men,' *FT Magazine*, November 11, 2006, 51.

[74] J. Derrida, *Limited Inc.* (Evanston: Northwestern University Press, 1988).

[75] This is to ignore the fact of where the act of writing took place in early modern England, which Fleming has argued was predominantly not in books but on walls and everyday material objects like pots. See Fleming, *Graffiti and the Writing Arts.*

[76] Fleming, *Graffiti and the Writing Arts.*

[77] Fleming, *Graffiti and the Writing Arts.*

[78] This is, of course, a sense of the world that has long been familiar to anthropologists and archaeologists in cultures where symbolism and daily life are intertwined in a network of entanglements which are both means of empowerment and dependencies, typified by Hodder's recent study of Catalhöyük with its sheer amount of elaborate wall art, stimulated by a lime-rich plaster that needed continual resurfacing and might be thought of as a prototype of the constantly refreshed screen. See Hodder, *Catalhöyük.*

[79] In fact, a few years ago four artists did construct a philosophical garden as an appendage to Erasmus herb garden in Anderlecht which is constructed on these premises, taking the text *The Religious Banquet* as the key.

[80] Fleming, *Graffiti and the Writing Arts*, 139–140.

[81] In 1916 the artists Vanessa Bell and Duncan Grant moved to Sussex with their unconventional household. Over the following half-century, Charleston became

the country meeting place for the group of artists, writers, and intellectuals known as Bloomsbury. Clive Bell, David Garnett, and Maynard Keynes lived at Charleston for considerable periods; Virginia and Leonard Woolf, E.M. Forster, Lytton Strachey, and Roger Fry were frequent visitors.

82 There could, of course, have been many other starting points, for example, the Palais Idéal of Ferdinand Cheval.

83 Herf, *The Jewish Enemy*, 274.

84 G. Bruno, 'Modernist ruins, filmic archaeologies,' in *A Free and Anonymous Monument*, J. Wilson and L. Wilson (Newcastle upon Tyne: Baltic, 2004) 7, 12.

85 L. Manovich, 'After Effects, or Velvet Revolution in Modern Culture, Part 1,' 2006, www.manovich.net

86 Frantz cited in Manovich, 'After Effects,' 8.

87 Such developments arise out of new practices combining with new outlets of expression (for example, most recently, YouTube).

88 A development that is only forced by developments in ubiquitous/pervasive computing.

89 Indeed, I think that it is quite possible to historicize the argument of Hoffmeyer and others that 'signs, not molecules, are the basic units in the study of life' (as cited in Wheeler, *The Whole Creature*, 123) making it into a symptom of the present.

90 Hansen, *New Philosophy for New Media*.

91 Indeed, one might argue that there is no longer a centre, only a halo.

92 For example, it would be possible to take a leaf from the art of landscape gardening again and suggest that what has become crucial is a knowledge of arrangement or disposition, of finding (search), which is currently going through a convulsion as the kinds of non-linear, non-discursive thinking and representation that I outlined in the previous section is brought back into play. Gardening may seem an odd metaphor to work from but I am sure that it fits the bill—passionate, sensuous, self-evolving, multi-sensory, synaesthetic, the favourite of Klee, the premier art of cultivation. See C. Tilley, 'The Sensory Dimensions of Gardening,' *The Senses and Society* 1 (2006): 311–330.

93 One thinks here of Artaud's influential 1938 essay, 'The Theatre and the Plague.' See J. Orr, *Panic Diaries: A Genealogy of Panic Disorder* (Durham: Duke University Press, 2006).

94 For me, the stakes are also personally high because I seem to have had to spend a good part of my career being belaboured by various hermetically inclined Marxists for following these lines of enquiry. See N. Smith, 'Another Revolution is Possible: Foucault, Ethics, and Politics,' *Environment and Planning D. Society and Space* (2007): 2

95 Orr, *Panic Diaries*, 18.

96 Orr, *Panic Diaries*, 8.

97 Foster, *Dances that Describe Themselves*.

98 'As dancers open their physicalized imaginations to entertain the possibility of any and all next actions, they also track the results of acting upon or rejecting those impulses. As viewers watch, going with the flow of events, they also critically engage with that going. Throughout the performance, both dancers and viewers ask themselves, what is going to happen next? And what difference will it make to this performance's significance?' Foster, *Dances that Describe Themselves*, 16.

99 Foster, *Dances that Describe Themselves*, 11.

100 Norman, S.J. *Locative Media and Instantiations of Theatrical Boundaries* (Cambridge: MIT Press, 2006).

101 J.T. Schnapp and M. Tiews, *Crowds* (Stanford: Stanford University Press, 2006).

102 S. Banes and A. Lepecki, eds., *The Senses in Performance* (New York: Routledge, 2006).

103 H. Lehmann, *Postdramatic Theatre* (London: Routledge, 2006); Zielinski, *Deep Time of the Media*.

104 The link to Whitehead's notion of prehension is clear.

105 S.J. Norman reference listed in N. Thrift personal notes.

106 For example, think of the ubiquity of Chernoff faces, which use facial features to represent data, or the growth of 'choreogenetics' which uses genetic algorithms to generate choreographic sequences. See Jones, *Sensorium*, 3; F. Lapointe, 'Choreogenetics: The Generation of Genetic Mutations and Selection' in *Proceedings of the Association for Computing Machinery Genetic and Evolutionary Computation Conference 2005* (New York: ACM, 2005), 366–369.

107 S. Ngai, *Ugly Feelings* (Cambridge: Harvard University Press, 2005); J. Gallop, ed., special issue on envy, *Women's Studies Quarterly* 34 no. 3–4 (2006).

108 H. Hampton, *Born in Flames: Termite Dreams, Dialectical Fairy Tales, and Pop Apocalypses* (Cambridge: Harvard University Press, 2007).

109 R. Barthes, *The Neutral* (New York: Columbia University Press, 2005).

110 *Ballettikka Internettikka: VolksNetBallet* took place on the same evening and at the same time as another people's festivity–the final match of the World Cup 2006, which also took place in Germany.

111 On another level, these performances are a part of general tendency to 'move the mutual implication of actors and spectators in the theatrical production of images into the centre.' Lehmann, *Postdramatic Theatre*, 186.

112 J. Pallasmaa, *The Eyes of the Skin* (Chichester: John Wiley, 2005).

Mobile Public Art and the Urban Screen

Maria Stukoff
Manchester Metropolitan University, Manchester Institute
for Research and Development in Art and Design

Art in the public realm is an increasingly influential part of Britain's buoyant urban renaissance shaping innovative perspectives of cultural practices in the built environment. With the proliferation of communication technologies and information processing ubiquitously embedded into civic architectures, the concept of the mobile 'networked city' is fast becoming a new art practice.

Mobile phones are now recognized as serious entertainment platforms: games and gameplay mechanics are employed by public broadcasting agencies, commercial retailers, galleries, and city planners to reach previously transient consumers and audiences via these devices. With the potential of the mobile screen to reach a global audience of billions at any given moment, these 'moving billboards'[1] can be considered the largest interactive screening platform in the public domain today. At the centre is the mobile and wirelessly connected individual, the 'phoneur,'[2] the modern equivalent of Charles Baudelaire's *flâneur*, seamlessly creating and enjoying leisure on the go.

This paper gives an overview into an investigation situating emergent new media art practices, particularly those utilizing mobile and wireless communication technologies, alongside a debate of 'art in the public realm' as a way to augment the built environment. This research was commissioned by the Manchester Digital Development Agency (MDDA) to enliven the city sphere as an interactive playground. It will also introduce the driving concepts behind *blu_box*, a Bluetooth-enabled mobile phone system that facilitates innovative forms of urban screen production, multiplayer gaming, and public broadcasting.

Introduction

Key to this research is an investigation into the relationship between these newly created wireless public artworks and traditional process-based art practices that encourage 'public intervention' and reflect 'conscious collaboration' as the artwork itself,[3] as advocated by the new genre public art movement.[4] These ideas inform my curiosity to engage and establish *blu_box* as a process-led, mediated event that cultivates 'collection action' and 'social interfaces' as suggested by Anne Galloway.[5] The *blu_box* project is concerned with how interactivity within technologically driven playgrounds or mobile-based entertainment zones can shift public attitudes

facing page

1 **Playful mobile interaction**
blu_box prototype testing
visualization, 2005

2 **Beyond the mobile**
Rethinking city infrastructures
for interactive play

CREATING FOR THE MULTI-PLATFORM CONTEXT

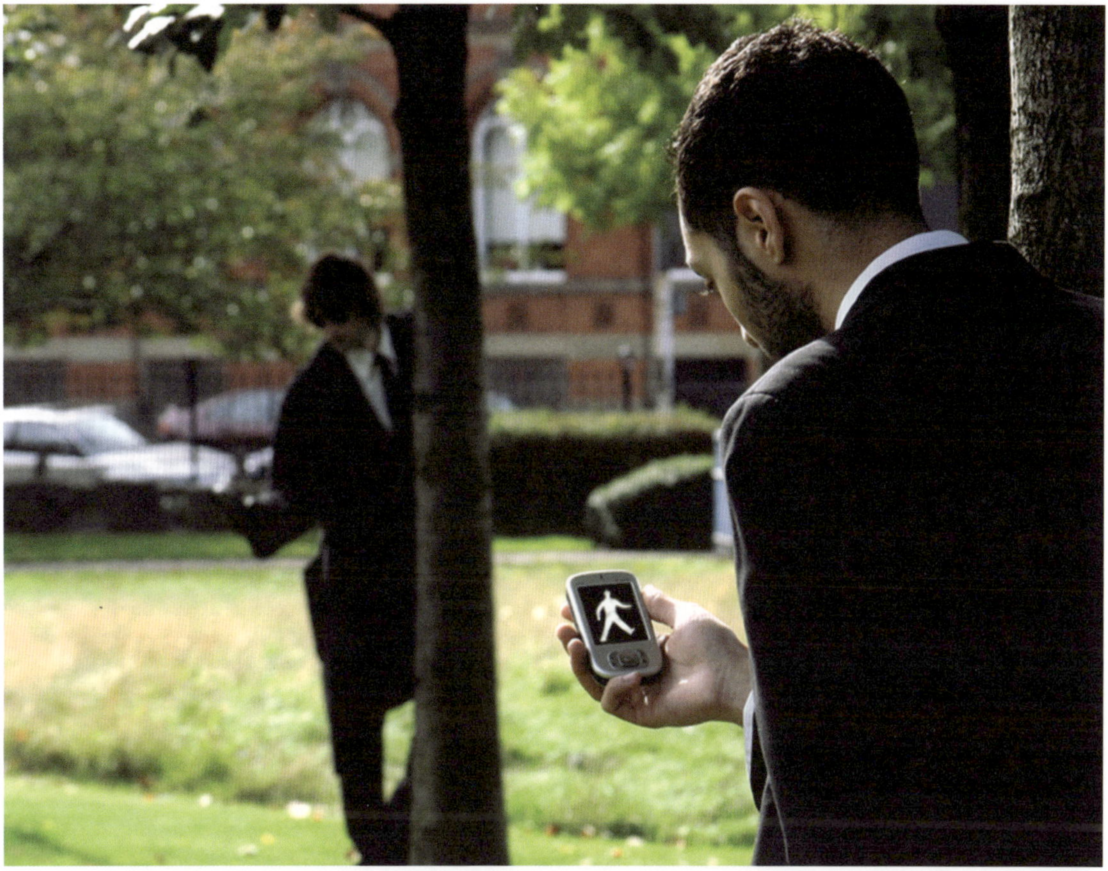

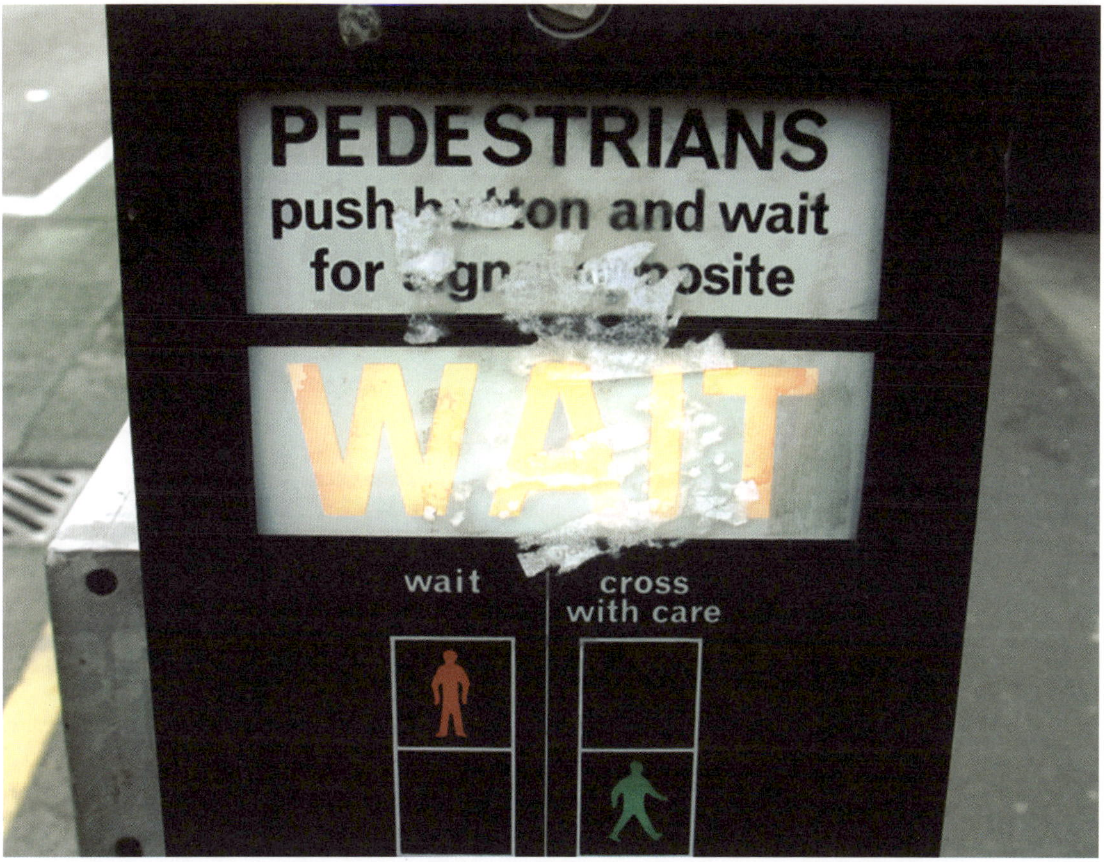

PEDESTRIANS
WAIT
wait
cross
with care

and relationships towards city areas that previously were considered only transitory. What happens to those areas once the public becomes an active ingredient or a live actor networking to the city space around them? *blu_box* aims to provide a better understanding of the significance of 'play' as 'art in the public realm' by staging innovative, collaborative, game-inspired events to examine how such new yet invisible city areas can assist city councils to construct dynamic environments fostering cultural identity in the city. *blu_box* builds on such influences and is fundamental to my research into 'socially dynamic' interaction between the public, urban spaces and the arts.[6]

3 *blu_box* creates ad-hoc social environments via mobile telephony

Fun and Games

In today's media-savvy society, digital games have become a global phenomenon no longer to be ignored as a minority pursuit. Digital entertainment encompassing video, music, and games has truly infiltrated our daily lives, informing social and cultural activity in the home, at work, or while on the move. Game environments are now often experienced as a digital activity, radically competing with traditional forms of physical playgrounds. Street games of yesteryear such as hula hoop or hide-and-seek are slowly fading away from public sight. Here I draw an important distinction between sport-oriented game activity, for example organized sports or skateboarding, as opposed to traditional play of chase, tag, or jump rope.

In his painting *Children's Games* (1560), Dutch/Flemish painter Pieter Brueghel (1525–1569) presents us with a stark reference point for the decline of public play in real, physical space. On careful inspection of the narrative depicted in the painting, one can see myriad forms of social interactions and play between children and adults in a city square, including games such as wedding imitations, tug-of-war, and see-saw.[7] Sadly, very few of these illustrated games can be witnessed in today's playgrounds, urban streets, or suburban backyards. Nowadays playgrounds are virtual landscapes experienced via a variety of different-sized electronic screens whereby playful interaction is experienced in the 'networked society' through a digital interface.[8] This electronically charged airspace becomes a special kind of 'responsive environment' where wireless communication technologies can map out and occupy lively and collaborative urban spaces.[9]

Thus, the city is re-emerging as a vibrant playground where artists are exploring the creative use of advanced 'urban social technologies' such as GPS, mobile telephony, and Wi-Fi.[10] This is echoed by Drew Hemment's sentiment that 'urban play is all about using the technology that surrounds us for creative ends.'[11]

Mobile Phone Bluetooth System

facing page
4 **Searching** for Bluetooth proximity readings, 2005

5 **Installing** the *blu_box* software, 2005

blu_box is a custom-built Bluetooth system which provides an innovative mobile platform for social play. The work is to be exhibited in a newly commissioned artwork linking mobile telephony with the BBC Big Screen in Liverpool. The system can consist of as many Bluetooth nodes as desired

Chat Now
RSSI
Close
Options
vodafone

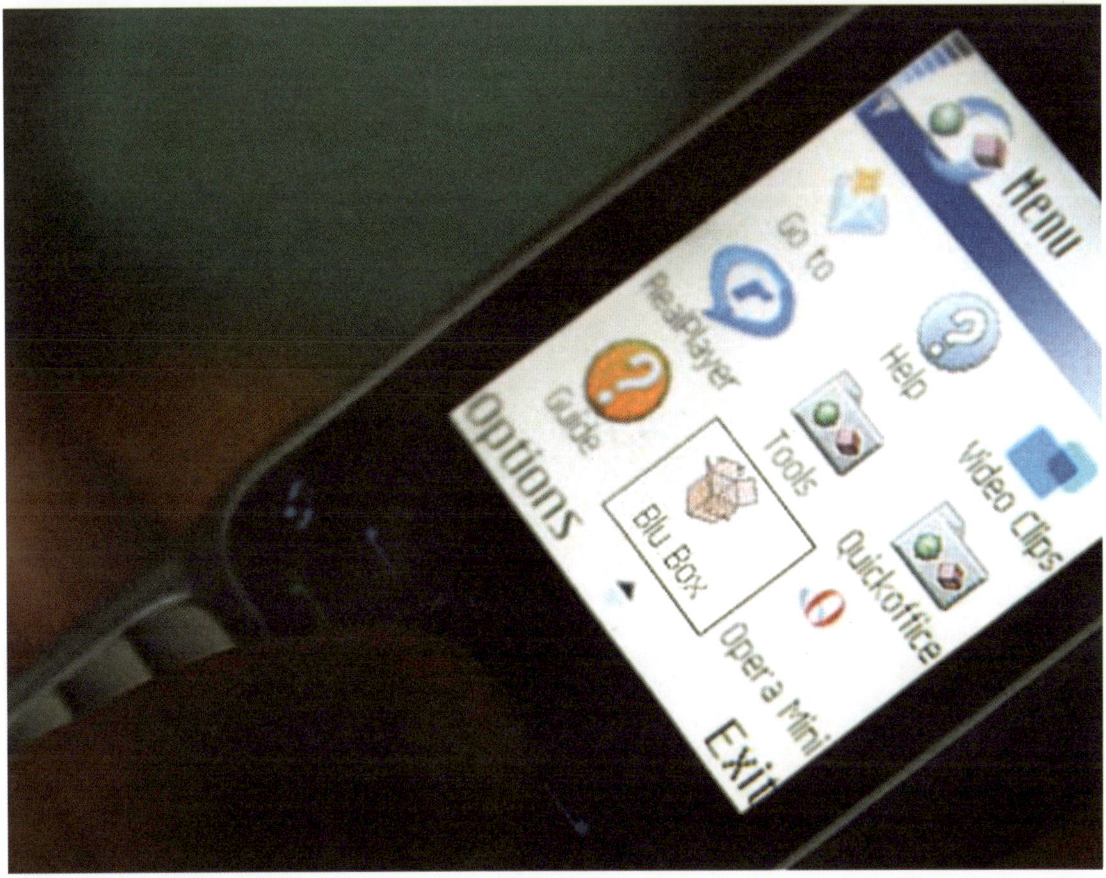
Menu
Go to
RealPlayer
Help
Guide
Tools
Video Clips
Options
Blu Box
Quickoffice
Opera Mini

to build a network across a city, along a street, or in a public arena. Nothing more than a standard Bluetooth-enabled mobile phone is required to interface with this system. In order to retain the qualities of 'spontaneity' and 'magic' generated by receiving a sudden text inviting users to play with their immediate surroundings, the decision was made at an early prototyping stage to move away from building an interface or stand-alone software package that needed to be downloaded by the participant prior to any interaction with a mobile game event. The driver for *blu_box* is the standard mobile phone Bluetooth mode that enables detection by proximity, triggering the receipt of text annotations about the zone, with clues on how to participate successfully in the mobile event at hand.

The mobile phone in public space becomes the interactive remote control to actively play with the sensitive and complex relationships that exist between the physical environment and virtual data we inhabit today. Thus, the individual walking the city has the capacity to influence, shape, and augment this media space both by themselves and through social interaction.[12] The searching Bluetooth nodes provide the mobile phone user strolling the cityscape with leisure on the go. This turns Charles Baudelaire's concept of the *flâneur*, the idea of wandering through the cityscape as a pursuit in its own right, into the modern equivalent of the 'phoneur.'[13]

The 'phoneur,' actively traversing the city sphere, is surveying the physical and the digital realm as *blu_box* is seamlessly searching, networking, and inhabiting a parallel social space. Robert Luke expresses a vision of the 'phoneur' within a commercial setting and e-consumer market, whereas I would like to consider the 'phoneur' within an artistic and cultural zone, where the invisible and digital data—pushed or pulled—will shape the creative information space linked to community groups, galleries, universities, urban magazines, etc. Such data will both interfere and interact with the city, disrupting yet animating public airspace.

It is these mobile radio waves that are waiting to be programmed, scheduled, and authored by the mobile public through free networking protocols such as Bluetooth, without subscribing to or paying mobile services requesting the exchange of 'personal information for the commercial privilege of using the networks.'[2] These enhanced city environments are the future 'sites for temporary relaxation,' creating transitory leisure zones, expanding on city activities such as drinking, eating, and other cultural events like theatre or cinema.[14]

Conclusion

The information-rich city 'is increasingly differentiated from previous urban forms by its extensive and interconnected networks for moving information.'[15] 'Mobility' as an art form offers artists innovative concepts about interfacing with audiences, ad-hoc communities, ephemeral networks, and public spaces. Perhaps in this sense, the notion of 'mysterious' rather than 'ubiquitous' networks refers to the ways that artists are designing interactive experiences in the public realm. Artists now have the ability to negotiate an art practice not only in physical environments or virtual spaces, but also have the potential to offer experiences that are simultaneously located in the real and digital worlds, where the notion of 'location-specific' brings with it new and open interpretations, as explored in *blu_box*. McKenzie Wark's prophetic sentiment seems to ring true: 'We no longer have roots, we have aerials.'[16]

Notes

[1] Lev Manovich, 'The Poetics of Augmented Space' (2002) www.manovich.net/texts_00.htm.

[2] Robert Luke, 'The Phoneur: Mobile Commerce and the Digital Pedagogies of the Wireless Web,' in *Communities of Difference: Culture, Language, Technology*, ed. Peter Trifonas (New York: Palgrave MacMillan, 2005), 185–204.

[3] Suzanne Lacy, ed. *Mapping The Terrain: New Genre Public Art* (Seattle: Bay Press, 1995), 11, 20.

[4] Miwon Kwon, *One Place After Another: Site-specific Art and Locational Identity* (Cambridge: MIT Press, 2004), 8.

[5] Anne Galloway, 'Playful Mobilities' (lecture, Ubiquitous Computing series, Emerson College, Boston, MA, January 26, 2005), http://institute.emerson.edu/floatingpoints/05/anne_galloway.php.

[6] Eric Kluitenberg 'Media Without an Audience,' <*nettime*>, (October 19, 2000), www.nettime.org/Lists-Archives/nettime-l-0010/msg00204.html.

[7] ArtyFactory, 'Perspective Drawing 14: Pieter Bruegel, *Children's Games*' www.artyfactory.com/perspective_drawing/perspective_14.htm.

[8] Manuel Castells, *The Rise of the Network Society: Information Age 1* (Malden: Blackwell, 1996).

[9] B.J. Novitski, 'Twenty-first Century Urbanism,' review of *E-topia: 'Urban Life, Jim—but not as we Know It'* by William Mitchell, *ArchitectureWeek.com*, www.architectureweek.com/2000/0802/culture_4-1.html

[10] Jens Pedersen and Anna Vallgarda, 'Viability of Urban Social Technologies' (Presentation, Sixth International Conference on Ubiquitous Computing, Nottingham, UK, September 7–10, 2004).

[11] Drew Hemment, 'The City as a Playground' (Program guide for Futuresonic Festival, Museum of Science & Industry, Manchester, UK, July 21–22, 2006).

[12] Michel de Certeau, 'Walking in the City' in *The Practice of Everyday Life*. trans. Steven Rendall (Berkeley: University of California Press, 1984).

[13] Luke, 'The Phoneur,'

[14] Michel Foucault, 'Of Other Spaces' (originally given as lecture, 1967) http://foucault.info/documents/heteroTopia/foucault.heteroTopia.en.html.

[15] Mitchell L. Moss and Anthony M. Townsend, 'How telecommunication systems are transforming urban spaces' (2002) www.mitchellmoss.com/books/transform_urban_spaces.pdf.

[16] McKenzie Wark, *Virtual Geography: Living With Global Media Events* (Bloomington: Indiana University Press, 1994).

Four Wheel Drift

Julie Andreyev
Emily Carr Institute of Art and Design

This paper presents two internationally performed projects produced under the title *Four Wheel Drift*: *FWDrift [remix]*, *VJFleet [redux]*, and the current research on *FWDrift *glisten)*. These projects examine urban culture through the use of 'expressive' cars fitted with audio and video recording technologies, sensors, interactive software, and media displays. During performances, audio-visions of the city are repositioned into the public and that which is private—the space of the car—becomes public and a tool for commentary about the city.

The development of the contemporary urban metropolis, the city as we know it today, has largely been determined and defined through the historical and spatial accommodation of vehicular transport. The car, in particular, has increasingly become the city's defining mobile feature, significantly contributing to its physical and social forms, and utilitarian and spatial functions. The complex and symbiotic relationship of the car and the space of the city, and its (not unproblematic) terms of coexistence, has seldom been the direct subject of art. The car as object, or the city as place, has achieved its own genres of representation in a variety of mediums.

Four Wheel Drift amalgamates aspects of 'low' and 'high' cultures: influences of car cultures, DJ cultures, LED signage, and mobile communications are fused with strategies from performance and installation art practices. The car can be seen to visually signify characteristics of city, such as social economies and hierarchies, and brand identification. In some car subcultures, modification of the car body provides a venue for individual expression to an otherwise mass-produced form. *Four Wheel Drift* quotes a racing strategy that causes a controlled, sideways slide while accelerating forward. 'Drifting' is now a popular subculture, with roots in the illegal auto-sport originating in Japan where drivers would drift along curved mountain roads. Drifting competitions, or 'battle drifts,' are judged events that examine the performer's speed, angle and style.

FWDrift [remix] and *VJFleet [redux]* use cars as mobile, interactive, experimental VJ/DJ platforms. Each time the projects are performed the interpretation of city space and experience is unique, and elements of perception are localized to each site. Creation takes place on the fly in a public setting rather than in the isolation of the studio. Influenced by artist group the Situationist International (1957–1972) the projects employ the tactic of the *dérive* (translates as 'drift') to cruise the city seeking out unexpected

1 **FWDrift [remix]** data gathering for performance at Pace Digital Gallery, New York, 2005

2 **FWDrift [remix]** performance for Elektra, Montreal, 2005

facing page
3 **FWDrift [remix]** data gathering for performance at Pace Digital Gallery, New York, 2005

4 **VJFleet [redux]** performance at SIGGRAPH, Boston, 2007

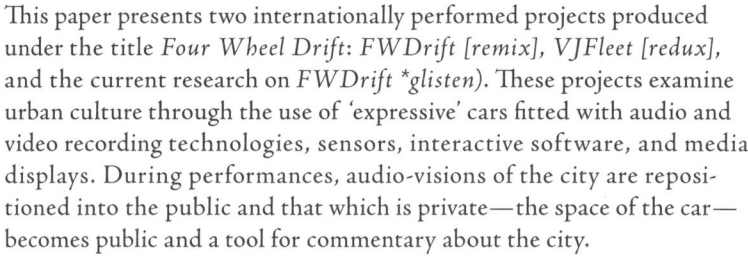

urban performance. The projects remove the car from everyday use as a functional means of transportation and reposition it as a provider of new forms of visual culture.

The projects involve modes of collaboration, such as those between the car and driver, between the production team and a local participant, and between the performers. The preparatory drive in the city relies on the participation of local passengers to direct a route and on the interaction between the car and driver. The passengers act as guides for the production of an audio and video archive used for the performances. Their local knowledge determines the specific visual mapping of the city and its vernacular highlights. As the cars cruise the city, conversations, the choice of music played on the car's stereo, directions from the local participants, and the team's responses are recorded. Cameras provide video imagery of the city manipulated by the interaction of the cars and drivers using VJ techniques. The movement and engine functioning of the cars, interpreted by sensors and software patches, create effects on the videos reflecting the choices made during the drive.

During a *FWDrift [remix]* performance, these video records become the visual playlist used by the software and VJ to generate panoramas. The audio archive of the drive is treated via software and the DJ to create a new, live musical soundscape. The car installed in the performance space functions as a pointer indicating the process of data gathering. During a *VJFleet [redux]* performance, the manipulated videos of driving routes are projected on panoramic screens on the cars visible to the street public. Software is used to remix the audio based on the sensor data of the drive: acceleration, braking, and turning. In this way, the physical experience of driving in the city is the motivating force behind the DJ and VJ manipulations of the materials.

*FWDrift *glisten)* proposes a medium that explores the city in an innovative encounter with real time; as a strategy this contributes comparatively as a 'drawing' or diagrammatic mapping of urban space through light and sound. The cars are fitted with sensors, software, and a custom-built skin of LEDs covering each car's body. As the cars cruise the streets, sensors and software interpret the drivers' interactions with the cars and the environment to generate moving light patterns and sound.

This project is informed by methods of car customization. Neon and small LED kits are used by enthusiasts to create light reflections on the pavement beneath or in the interior of cars. The aesthetics of custom paint suggest specific symbolic meanings, such as speed, status, and utility. *FWDrift *glisten)* borrows from these practices by creating light effects on the body of the cars as 'animated custom paint.' Normally, in the private space of a car, the driver and passengers are consumers of music. *FWDrift *glisten)* examines these areas of consumption by positioning the drivers and the cars as *creators* in a public space. Here, the car is seen as an instrument for content creation in an improvisational performance.

5 **FWDrift [remix]** data gathering for performance at Pace Digital Gallery, New York, 2005

6 **Simulation** of *FWDrift *glisten)* performance

By using the car as a platform for experimentation, a rethinking of the artist's studio space occurs. In this case, the artist's space is mobile and subject to influences of the site. The cars, as mobile interactive platforms, function as information-gathering and dissemination devices that collect and distribute information about the city. The performance can be referenced in terms of Michel de Certeau's theory of 'tactics' within everyday life.[1] Tactics depend on temporal moments that are 'seized' and manipulated into 'opportunities.' The drivers and cars are positioned in this urban performance as comparative 'tactical agents' that make use of spontaneous events to transform and rethink urban experience.

7 *VJFleet [redux]* performance at SIGGRAPH, Boston, 2007

Note

[1] Michel de Certeau, *The Practice of Everyday Life* (Berkeley: University of California Press, 1984).

Storytelling Goes Mobile

Shawn Micallef
[murmur]

[murmur] is a location-based mobile phone documentary project that is active in Toronto, Montreal, Vancouver, Calgary, San Jose, and Edinburgh. The project records stories about particular locations by people with a connection to that spot. Some stories are purely anecdotal—'I fell in love on this park bench'—to stories that describe the wider social, civic, and political history of that spot and surrounding location. Some storytellers are able to blend both their own experience and historic details at a location, but all are told from a personal perspective. [murmur] makes every effort to record storytellers in situ, as the geography reminds storytellers of forgotten details and puts them at ease, encouraging a more natural conversational tone.

Once stories are recorded, [murmur] installs a sign at that location with a phone number and unique location code that anyone with a mobile phone can call and listen to stories about that spot. When heard, it's as if storyteller and listener are out for a casual stroll through the neighbourhood.

At their core, [murmur] stories are able to convey the emotional attachment people have to places, and let them explain why this location is important. [murmur] is also a way of distributing stories and histories that are not the 'official' history of a place or city, providing a way for overlooked voices and memories to be distributed, democratizing who controls a city's official narrative

[murmur] is a psychogeographic experience, allowing people to access that emotional, human layer of memory about a place via mobile technology. And, though a project employing technology, with [murmur] technology must always be at the service of the experience, facilitating a direct connection between the listener and the storyteller. As such, by using personal mobile devices, listeners do not have to learn how to use a new piece of equipment, allowing the familiar technology to disappear.

1 **A [murmur] sign** on Spadina Avenue, Toronto

facing page
2 **A [murmur] sign** in Kensington Market, Toronto

Shorts In Motion

Judy Gladstone
Bravo!FACT (Foundation to Assist Canadian Talent)

Launched in 2005 by CTVglobemedia's Bravo!FACT (Foundation to Assist Canadian Talent) and the National Film Board of Canada (NFB), the *Shorts In Motion* (SIM) project has captured the attention of the wireless industry, academia, media, and the public, both locally and worldwide.

Commissioning shorts for viewing on multiple platforms was initiated following a presentation by a British Hewlett-Packard researcher at the Canadian Women in Communications New Media Career Accelerator Program at the Banff Centre in the winter of 2004. The presentation referred to video cellphones being used in a research study in the UK. Realizing that video cellphones would be entering the North American marketplace, Bravo!FACT approached the NFB and suggested they partner on a project to give Canadian telecom companies the opportunity to provide original Canadian content.

The project evolved into the 2005 edition of *Shorts In Motion* involving filmmakers Mark McKinney, Don McKellar, Sudz Sutherland, and Sook-Yin Lee, and producers Daniel Iron of Foundry Films and Jennifer Weiss and Simone Urdl of Film Factory. While neither the filmmakers nor the producers had seen or heard of video cellphones when approached in the winter of 2004/2005, they were intrigued by the challenge of creating a short that could be viewed and enjoyed on all sizes of screens (theatrical, television, handheld device, and computer). The filmmakers were each lent a video-enabled cellphone as inspiration while they shot. Don McKellar delivered two shorts, *Phone Call From Imaginary Girlfriends: Ankara* and *Phone Call From Imaginary Girlfriends: Istanbul*; his works were the first films in the world completely shot on a video cellphone (edited on a computer) and blown up to 35 mm for theatrical release. The project captured the imagination of the Canadian public and media, and the shorts by McKellar and McKinney received the only two Canadian nominations for the Mobile TV Awards at the MIPCOM Conference in Cannes in 2005.

The 2006 *SIM* project was larger in scope.[1] While the 2005 *Shorts In Motion* films had no common theme, had an average running time of four minutes, and involved four filmmakers, the 2006 edition was quite different. Ten filmmakers were commissioned, nine from across Canada, including Ann-Marie Fleming, Trent Carlson, Guy Maddin, filmmaking duo Adam Brodie and Dave Derewlany, Jenn Goodwin, Mark McKinney, Theodore Ushev, Denis Villeneuve, and Anita McGee, with the tenth short providing Isabella Rossellini

facing page
1 **Strip Show** directed by Adam Brodie and Dave Derewlany

2 **Sou** directed by Theodore Ushev

[1] **For the 2006 *SIM* project,** the private-public sector partnership of Bravo!FACT and the NFB was repeated, with the production company Marblemedia being the third creative partner. New funding parties (Telefilm's Canada New Media Fund, Ontario Media Development Corporation's Interactive Digital Media Fund, Ericsson Canada and the Sundance Channel) also contributed to the 2006 edition.

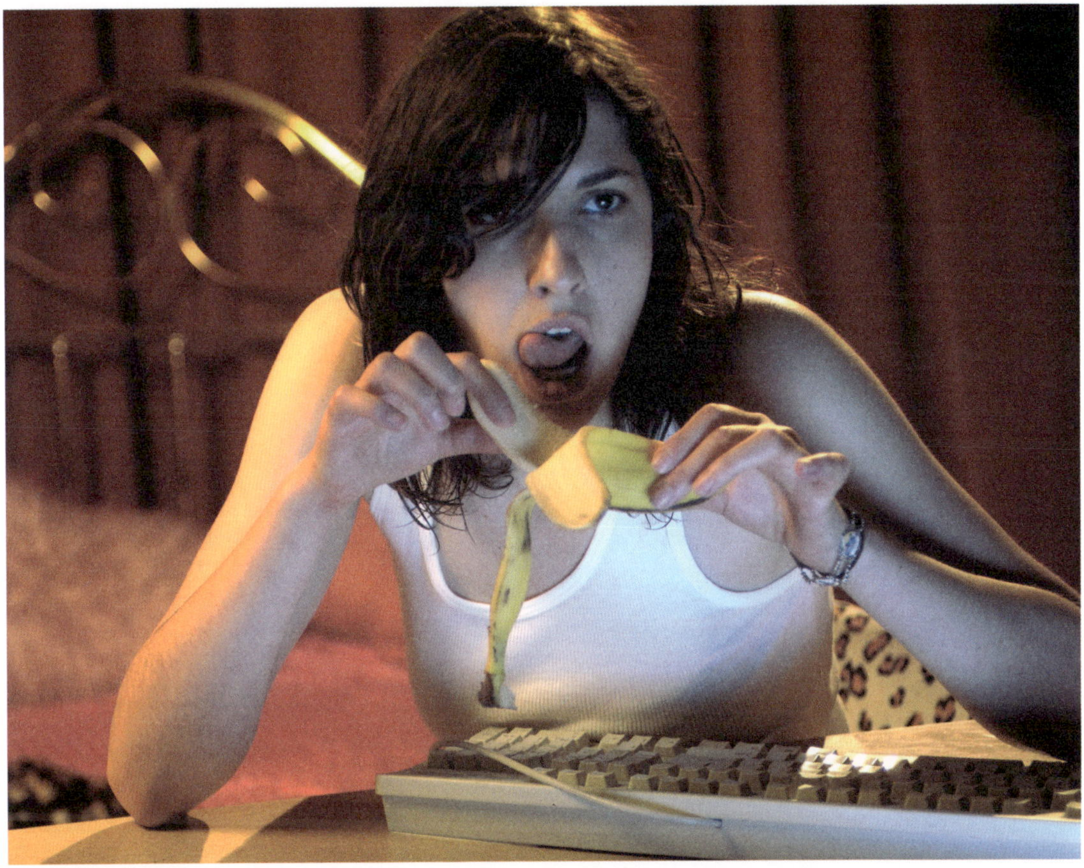

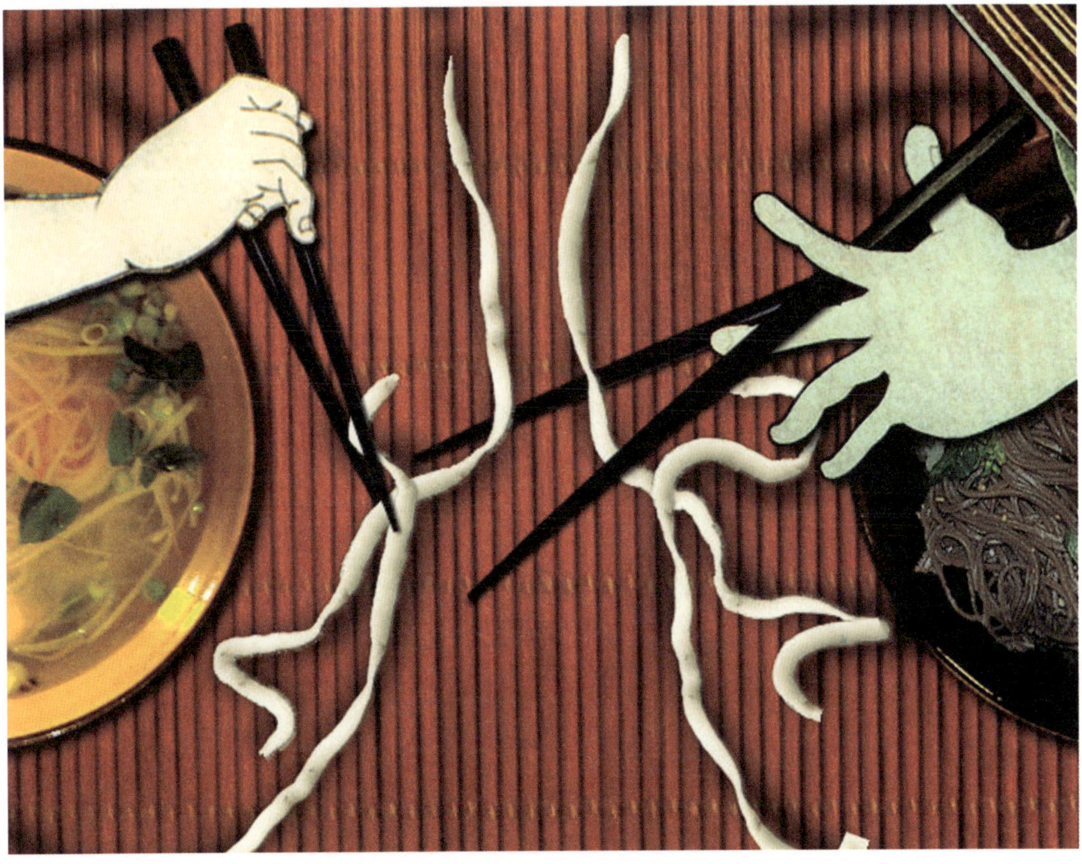

2 **Shorts In Motion: The Art of Seduction** has been honoured with the prestigious 2007 Global Mobile Award for excellence in the development of new video content for mobile devices and was also nominated for a 2007 International Interactive Emmy Award for Best Interactive Program. It won the Banff Rockie award for Best Mobile Program Enhancement at the 2007 Banff Television Festival.

3 **John McKay,** 'Winnipeg's Guy Maddin Pleased with his Long and Short Film Fest,' *Ottawa Citizen*, September 17, 2006.

4 **John McKay,** 'Movies Made for and by Cellphones a Toronto International Film Festival First,' *Montreal Gazette*, September 14, 2006.

facing page
3 **Nude Caboose**
directed by Guy Maddin

with her directorial debut. Each of the filmmakers was given artistic freedom to explore the theme of 'the art of seduction,' with each short having a running time of two minutes.[2]

Shorts In Motion constitutes a significant development in the mobile video industry. These shorts represent the first real content, with high production values, commissioned for mobile devices. Content for video-enabled mobile devices is traditionally repurposed from existing movies, web content, and television programs or is intended to market a product (a music video, for example). *SIM* was among the first projects worldwide to commission original work created for multiple screens. The content is fluid, as likely to be viewed on a handheld device as to be seen while channel-surfing. The project also responds to the lack of original content being developed for the mature viewer. In North America, watching video on a mobile device is considered a teenage phenomenon; however, existing content neglects a very important demographic, the adult user who, often due to their business environment, has access to the newest handheld technologies. It is this demographic that is looking for a unique cultural experience with their mobile devices.

The challenge of creating compelling content for multiple screens also provided an artistic test for the *Shorts In Motion* filmmakers: how to create a film that would look as good on a tiny handheld screen as it would on 35 mm at a festival. In addition, some of the filmmakers pushed the craft further by experimenting with the cellphone camera to shoot their project. Director Guy Maddin uses vintage cameras and films as part of his normal filmmaking process. In an interview with Canadian Press writer John McKay, Maddin says of his short *Nude Caboose* that 'it looks like the world's largest, filthiest aquarium'— and he notes this with pride, not disappointment.[3] Don McKellar, in another McKay article, points out the artistic potential of the medium: 'At this moment at least, the technology of shooting with a cellphone has a kind of almost old-fashioned rawness to it that is emotional and moving.'[4]

While *SIM* gave filmmakers an opportunity to experiment, it has also been successful at providing avenues of exposure for short films. As technology, and the ability to record, upload, and distribute content, reaches the hands of home users, audiences have begun to fragment. Traditional methods of distribution and viewing are no longer as viable as they once were. Perhaps one of the most overlooked but important aspects of mobile technologies and projects like *SIM* are their ability to democratize art. The adaptability of cellphone technology to the short film genre means that audiences, theoretically, have easy access to this content. They don't have to show up to a film festival; they don't have to subscribe to cable television. These artistic gems can be accessed anywhere there is a cellphone signal. In turn, this model of distribution has the potential to provide exposure for Canadian filmmakers in ways the conventional broadcast system and film industry have not, thus breathing new life into the short film genre.

Mobile Text Messages as Part of an Interactive Television Drama

Leena Saarinen
Helsinki University of Art and Design

Accidental Lovers (originally known as *Sydän kierroksella*, a 2006 Finnish production directed by Mika Tuomola)[1] is an interactive musical TV comedy that explores variations of a love relationship between sixty-one-year-old cabaret singer Juulia (Kristiina Elstelä) and thirty-years-younger pop star Roope (Lorenz Backman).

Accidental Lovers introduced a new interactive story format and genre for television where viewers were able affect the unfolding drama, including turns and outcomes of the plot, by sending mobile text messages (also known as SMS messages) to the broadcast. Each message was scanned for keywords that could cause particular types of thoughts (manifested as voiceover audio clips) in the main characters. As the keyword data accumulated in text message chat, the characters' relationship changed.

Broadcast Results

Accidental Lovers was broadcast on Finland's YLE TV1 channel in four episodes between December 28, 2006, and January 5, 2007, featuring a total of twelve story variations. While a single story variation reached about 100,000 viewers, at its highest altogether the series reached 1,085,000 viewers. Audience ratings were slightly below the average ratings on YLE TV1 channel on Wednesday and Friday evenings, but considering the novelty of the format the broadcaster was satisfied with the ratings. Interestingly, the viewer profiles regarding gender and age resonate with those of the drama's main characters: the largest audience groups were twenty-two- to fourty-four-year-old males (25 percent) followed by forty-five- to sixty-four-year-old males (20 percent), women over sixty-five (18 percent), and women from forty-five to sixty-four years of age (16 percent).[2]

Almost 3,000 mobile text messages were sent to the broadcasts from 1,300 different numbers. The messages that arrived to the system can be categorized according to message content. The following are categories that clearly stood out in order of frequency:

Advice/command to a character: 'Juulia be tender to Roope, don't leave him alone to the storms of life.' Opinion about character: 'Roope is young and hasty. He cannot concentrate. He has no clue yet. Juulia will suffer.' Theme-related thought (love, age difference, death): 'Love is a promise of a grand illusion!' Fantasy scene (fan fiction): 'Cat runs away and Juulia goes looking for it but Roope finds it first and Juulia gets lost in the city then Roope goes to Juulia's place but Juulia is not at home.' User feedback on form

Sidebar notes:

1 **Juulia (Kristiina Elstelä)** and Roope (Lorenz Backman) in *Accidental Lovers*. User Inter-face information for the viewer: 'Your message will give Juulia or Roope a thought that affects their heart. Send your SMS to Juulia or Roope to number 17239.' The text message: 'Go for it crazy Roope! You'll die in the end anyway!'

1 **The concept** and script for *Accidental Lovers* is originated and written by Leena Saarinen and Mika Tuomola

2 **Television** audience measure-ments were conducted by panel research company Finnpanel Oy

of the show (interactivity, genre, user interface): 'Program texts are so small and difficult to read that an older person could not participate even if she wanted.' Frustration on not getting messages to the broadcast: 'Now I will switch channels. Boring that nothing pleases you!' Reply to a previously received return message: When users got a return message from either Juulia or Roope some of them replied, shifting to quite an intimate mode with the character. This kind of content was usually such that it did not make sense to accept for the broadcast, so the users received another return message: 'The message was not intended for your information. The intention was that Roope should examine his inner self. Women can be friends but only a man can be beloved.' Off-topic: 'Happy New Year!'

The majority of messages were 'red hearts' referring to the happy love relationship instead of the 'blue hearts' for the idea that the main characters should not fall for each other. This was a minor challenge in moderating because we wanted to present all kinds of plot variations through the course of the broadcasts. Overall, the viewers' text message content was really fantastic. It was deeply involved with the drama, the characters, and the main themes.

Conclusions

Accidental Lovers demonstrated that there is strong potential to develop engaging interactive content for a TV audience. Some of the viewers got activated and sent messages that were extremely well connected to the show's themes and its characters' destinies. There were some expressions of disappointment from viewers that sent many messages but never got their message on the TV screen. Still, it can be claimed that the return messages sent by the main characters were clearly supporting the interactivity because some viewers reacted to these too by sending replies to the return messages. For future use the return message system should be developed towards more conversational interaction where users can continue text message chat with a fictional character.

Broadcasts of *Accidental Lovers* achieved a satisfying number of viewers and extremely good user-generated content (text messages). The viewers' willingness to discuss the drama's themes and its fictional characters' destinies was hooked to the story world in a way that encourages further investigation into design for interactive stories, where user-generated content is an important part of the ongoing narrative.

Mobility and the Identity Continuum

Nathon Gunn
Bitcasters

One of the more interesting frontiers in mobility lies at the intersection of two important online phenomena: social networks and multiplayer games. At this intersection lies a continuum of community, play, economy, environments, and above all, identity.

Although this range of experience exists in many online arenas, including commercial examples like E*TRADE and eBay, things start to look even more interesting as we mix in social sharing and community features such as del.icio.us.[1] This becomes even more intriguing when we put this mixture into the palm of our hand and start moving through a physical landscape and around other people. Alternate identities and virtual realities skew this even further.

From its position as a commercial-based research and 'real life' laboratory environment, Bitcasters offers validity to observations and conclusions regarding shortened product life cycles, faster product delivery, and real-time monitoring, with important research and implementation roles in several innovations including a precursor to YouTube, an early secure digital music format, and the first hybrid TV/web chat.

Today much research and commercial development is based on the identity continuum concept: the ability for a person's virtual identity to transcend fluidly across multiple, disparate applications and media. A particular area of exploration in this vein is a series of commercial, multi-platform entertainment applications where users create, manage, and nurture virtual creatures. These creatures exist in a persistent state in a shared database through which other mobile users, web visitors, and game players interact.

At its most basic level, this combination of features does exist elsewhere, most notably with Neopets. The concept allows children to create and nurture their own virtual pets and interact with a social network of millions of other pet owners through games, chatting, and trading virtual pet items such as clothing and accessories. As of June 2006, the Neopets continuum extends to mobile space as well.[2]

While it may tempting to write off the Neopets model as (literally) child's play, it shows how powerful mobility, gaming, and social networks can be when they are extended into other applications. Moreover, the entire games medium is evolving to the larger consumer population as a whole. Game players are quickly increasing in age, professionalism, and sophistication. Corporations from Microsoft to Reuters have invested in

[1] **E*TRADE**
(www.etrade.com)
is a financial services company based in New York City, the major business of which is an online discount stock brokerage serving self-directed investors

eBay Inc.
(www.ebay.ca)
is an American Internet company that manages an online auction and shopping website where people and businesses buy and sell goods and services worldwide

del.icio.us
(http://del.icio.us/about)
is a social bookmarking website. The primary use of del.icio.us is to store bookmarks online, which allows access to the same bookmarks from any computer and the ability to add bookmarks from anywhere.

[2] **Neopets** is located at www.neopets.com.

facing page
1 *Hollywood Tycoon Mobile*
The *Studio Game* lets players build their own movie studio empire including (as shown) western movie sets, writers huts, actors trailers, and more

2 *Hollywood Tycoon Mobile*
Once you've chosen your script and hired your star actors and directors, it's time to build a set (in this case, for science fiction movies)

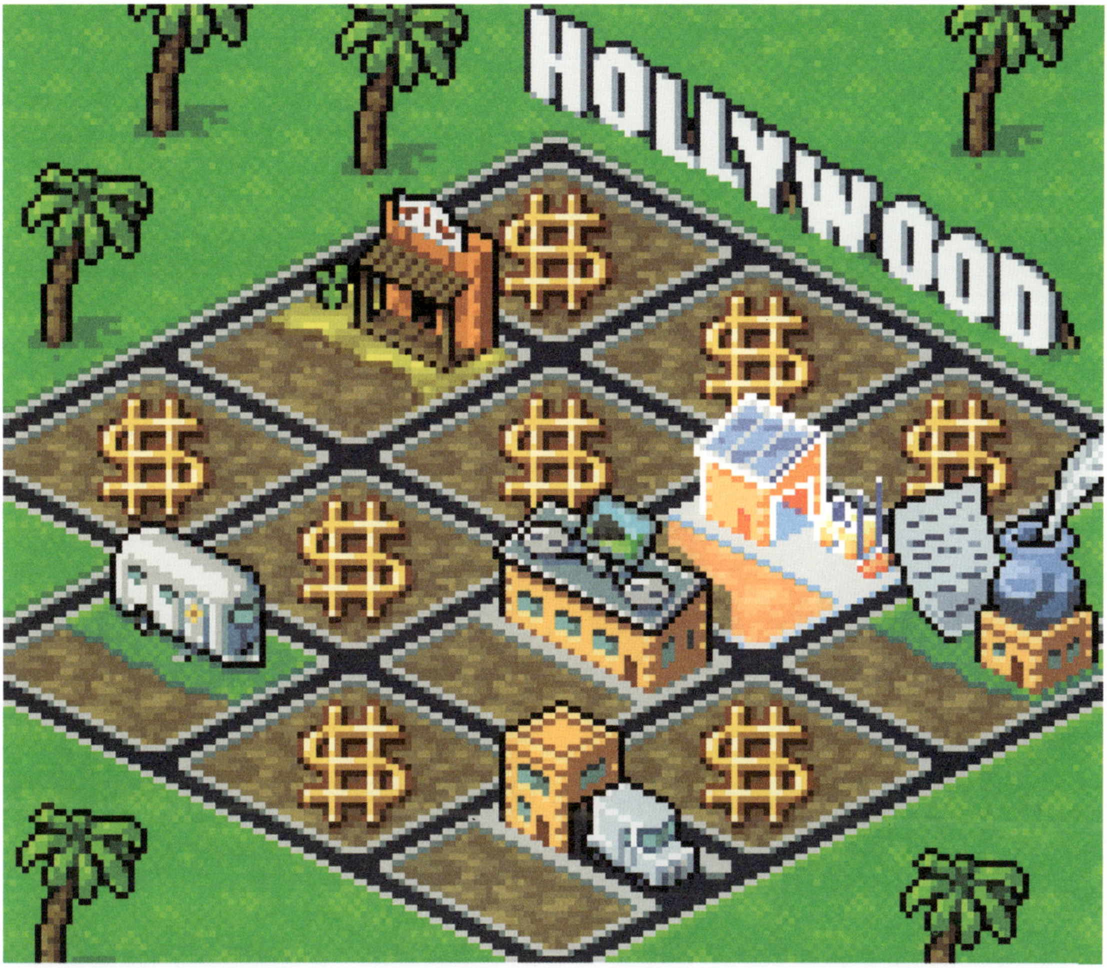
HOLLYWOOD

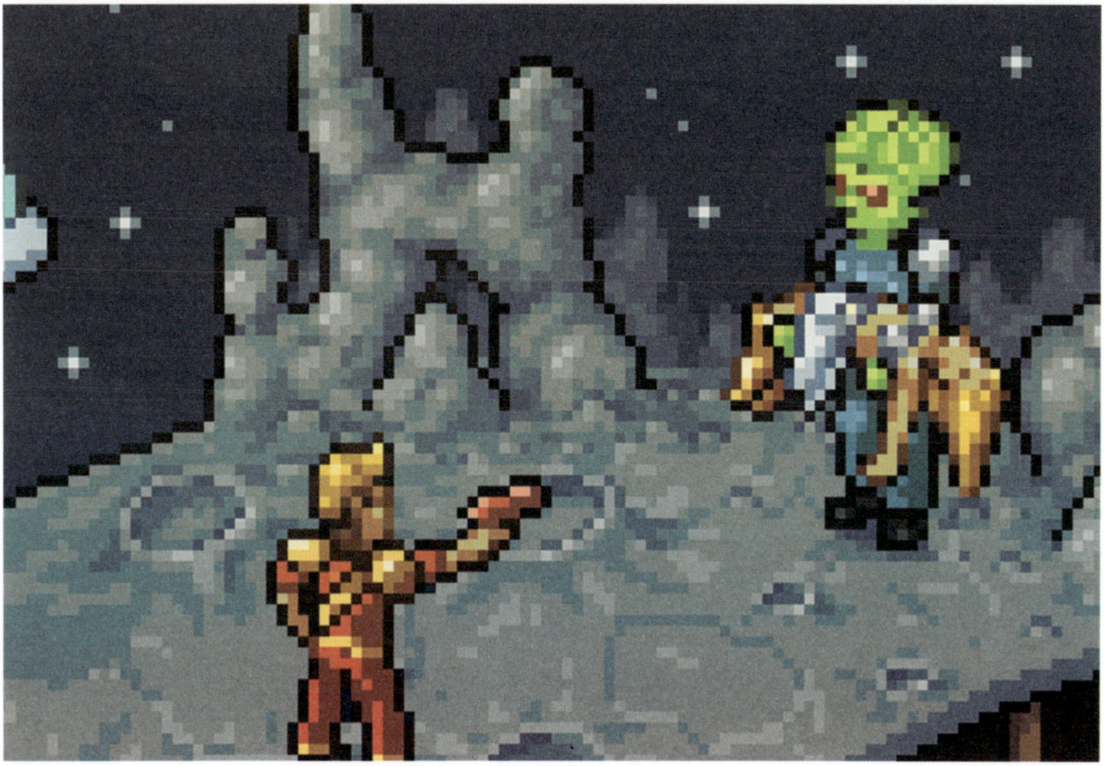

3 **MMOG** (a massive multiplayer online game) is a computer game that is capable of supporting hundreds or thousands of players simultaneously, which is played on the Internet (http://en.wikipedia.org/wiki/Massively_multiplayer_online_game)

Second Life (http://secondlife.com) is an Internet-based virtual world, a 3-D digital space imagined, created, and owned by its residents

3 **Hollywood Tycoon Mobile** The *Agent Game* allows players to hire and manage a stable of actors and directors, and build their value through trade and interaction with other players

facing page
4 **Hollywood Tycoon Studio** desktop game beta / pre-release

virtual events and even offices in MMOG environments like *Second Life*,[3] a testament to the validity of these games' virtual economies. What's more, so-called serious games are proving their value in teaching, training, and even combating addiction.

In short, the line between fact and fiction is quickly blurring, and the increasing ability to switch from 'game mode' to 'real world' creates some exciting new social, entertainment, and commercial opportunities. Consider just two of the possibilities, both from the research and commercial perspectives, of extending the continuum of identity, commerce, and communications across the real and virtual worlds: establishing a single virtual identity that represents you online, in-game, in email and instant messenger communications, and even on call display, and using currency earned through game play, item trading, or successful social network connections to make physical purchases using online or mobile-commerce payments.

To demonstrate some of these principles, witness *Hollywood Tycoon*, an example of 'real-world commercial laboratory' research and a unique experiment in combining two basic multiplayer game models. In the *Studio Game*, available as a downloadable 'casual' computer game for both desktop and mobile, players build their own movie studio empire, creating sets, hiring actors, and shooting movies. In *Agent Game*, a purely online (desktop and mobile browsers) game, players are agents to their own actors and directors, shopping them to studio executives, trading them with other agents, and making the right investments (publicity, training, etc.) to make their actors and directors Hollywood legends. These two games represent advances in online entertainment, but *Hollywood Tycoon*'s innovation is magnified into one global economic, social, and entertainment environment. Studio players aren't simply playing against a hidden algorithm, but using the player-generated content from the *Agent Game* to build their empire. Similarly, *Agent Game* players aren't simply grooming a virtual pet, but building a stable of talent that ties directly into the *Studio Game* and serves as a currency exchange as agents invest in and trade their talented stars. The addition of mobility only makes the experience more visceral.

Looking forward, opportunities abound for games such as *Hollywood Tycoon* to extend the identity continuum principle further. Consider the possibility of real actors or scriptwriters using the game to shop their wares virtually before knocking on real studio doors; studios could enter the fray, with the game serving as a proving ground for scripts and actors they otherwise wouldn't have time to assess.

On a practical level, moving items, currency, identity, and other bits of data from device to device while preserving security, fluidity, and connectivity is a technical and logistical challenge. But as these problems are solved—a matter of implementation and adoption, for technologically the solutions already exist—the possibilities will become realities, and the notion and importance of the identity continuum will become clearer.

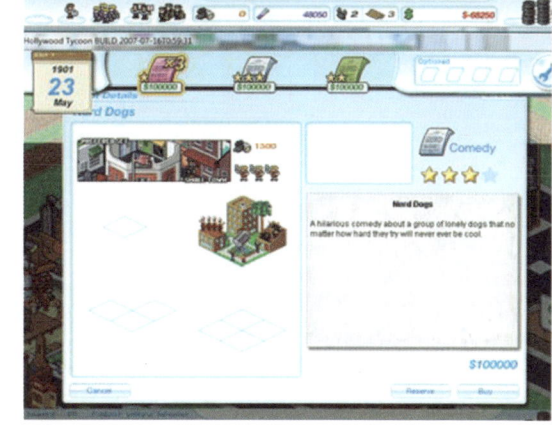

Pervasive and
Social Computing

The conjoined theme of pervasive and social computing offered the opportunity to involve the public in the Mobile Nation conference.

The extension of mobile computing into the realm of pervasive computing includes the embedding of computer technologies and sensors into spaces and artifacts, in combination with artificial intelligence software that can, in turn, respond to and 'reason' about human actions and behaviours. In fact, mobile experiences can move beyond text, sound, and image. In response to the needs and desires of the user, spaces will become highly intuitive. We can think of architecture's and outdoor space's 'software'—the ephemeral sounds, smells, images, temperatures, and even social relations that surround them—and program the way these spaces interact with their community of end-users.

In the area of wearable computing, Mobile Nation further investigated the potential of devices to deliver a rich variety of engaging user experiences that enhance everyday activities and situations through context-sensitive media and interaction. The potential for new forms of mobile expression are only beginning to emerge and deserve attention from both research communities and industry interests.

Context, Content, and Community
inventing the future of mobile media

Marc Davis
Yahoo! Inc.

With the advent of the camera phone as a ubiquitous platform for media production, sharing, and use, we have the opportunity to reinvent media participation on the personal, social, and global levels. We can now gather and correlate media metadata about the spatiotemporal context and social community of media capture and use it to enable people around the world to create, describe, find, share, and remix media content.[1] To do so, we are engaged in the iterative design, development, and analysis of large-scale 'sociotechnical' systems that will ultimately connect billions of humans, computational devices, and media assets into a global processing network. The design of these sociotechnical systems for mobile media participation requires us to rethink the core assumptions and boundaries of computer science, information science, social science, media studies, and design, as well as to create new processes and organizations for interdisciplinary collaboration.

The key intellectual shift reframes technological challenges for mobile media within social and humanistic understandings of information, communication, context, memory, and identity, and iterates these humanistic and social understandings within computational theory and practice.[2] Computational models of information processing have largely not availed themselves of humanistic and social scientific models of how information is constructed, interpreted, shared, accessed, and reused by embodied humans in spatial, temporal, social, and topical contexts. The technological challenge of analyzing, routing, and retrieving media content becomes solvable in new ways when informed by an understanding of what information is and where and how it is processed by human beings in the world. Simply put, when technologists come to understand that the meaning of the data we process is not in the data itself, but elsewhere, then better technologies can be created. Where is the meaning in data? Meaning is a process whose traces can be read in the patterns of where and when and by whom data is created, shared, and recombined. By interacting with the affordances of physical places, temporal events, and social and cultural structures and norms, human beings create patterns of activity, the analysis of which can help us predict what people might create, want, and share in a given spatial, temporal, social, and topical context *(fig.1)*. These hitherto invisible and ephemeral patterns of human attention and activity in the world and with each other can now be captured by our cellphones as mobile media metadata and made available to sociotechnical analysis

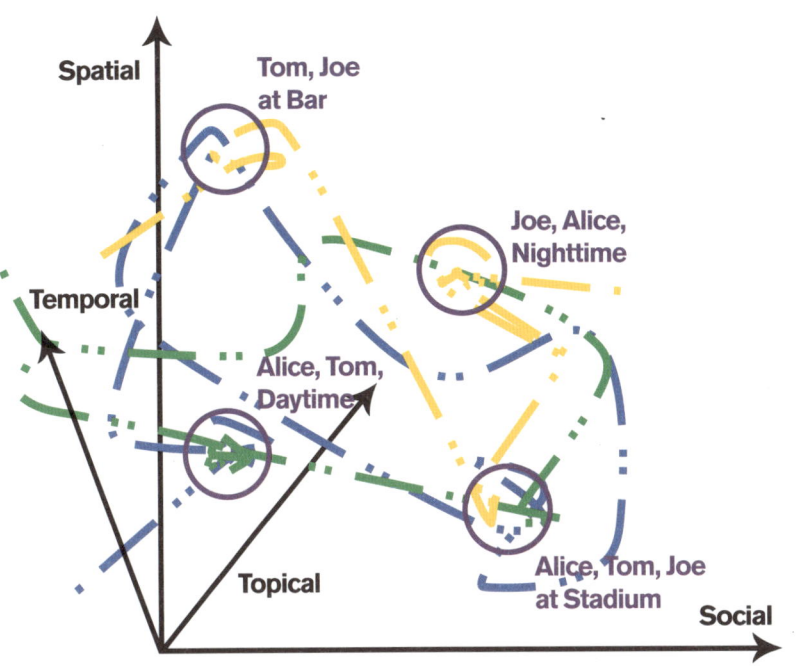

Exercise 'What' can we learn about people from their daily activities?

Hint 1 You're given Where, When, Who (some explicit, some implicit).

Hint 2 Three of the four Ws give us considerable contextual info, enough to infer 'What.'

Answer We surmise that Joe and Alice are married. Alice and Tom work together, all three went to a performance together, and Tom and Joe are friends who had a drink together afterward.

Spatial

Temporal

Topical

Social

Tom, Joe at Bar

Joe, Alice, Nighttime

Alice, Tom, Daytime

Alice, Tom, Joe at Stadium

Alice	— — ·
Tom	— — ·
Joe	— — ·

Thanks to Carter Trout for greatly enhancing this illustration.

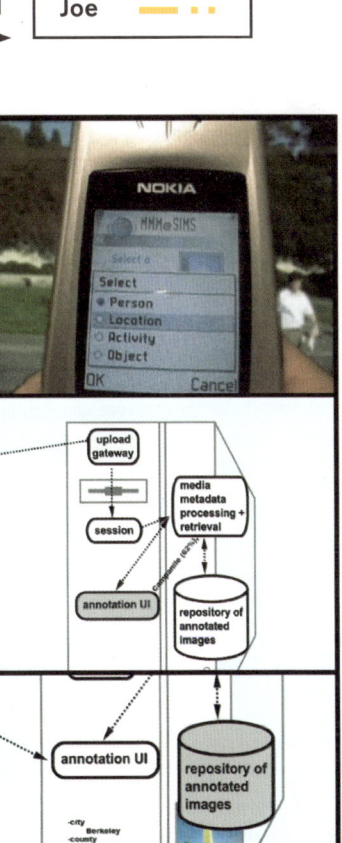

3 **Phone screenshots from MMM2** (Mobile Media Metadata 2) prototype for mobile media (photo and video) upload; automatic metadata capture and upload; and context-aware suggestion of likely sharing recipients for captured media

4 **Web screenshot from MMM2** web interface. Photos with star icon have automatic location metadata from GPS and photos with person icon have copresent people metadata automatically sensed via Bluetooth

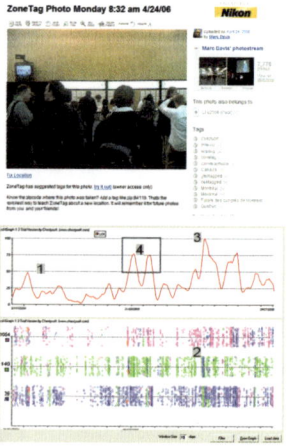

5 **Screenshot of ZoneTag** photo on Flickr with accompanying tags including tags selected by user from ZoneTag's context-aware suggested tags, as well as automatically added and inferred tags for location from CellID and GPS

6 **PhotoLOI** (Level of Interest) diagram visualizing socio-temporal patterns in three users' photos in MMM2. Upper graph de-picts total number of photos taken over time, lower graph depicts photos taken by each user—peaks in the upper graph indicate when more users took more photos at the same time.

and design.[3] The 'where' of human activity and attention is increasingly available to researchers and designers through various mobile sensors (Wi-Fi, Cell ID, GPS, Bluetooth, etc.) and the correlation of these data with geographic information databases and maps.[4] The 'when' is available through the network time of cellphones and can be connected to online calendar and event databases and modelled in terms of periodic temporal patterns.[5] The 'who' of mobile media metadata is available through the authenticated identity of mobile phones, the address books and communication logs of phone users, and 'copresence' sensed through Bluetooth and other means,[6] and may be correlated with various web-based identity sources and social networks. Integrating and correlating this spatial, temporal, and social metadata with the media content people create and use makes possible new ways of understanding, modelling, and visualizing human behaviour and attention in the world, helping initiate an emerging 'computational social science' of mobile society.

Over the past five years, first at Garage Cinema Research in the UC Berkeley School of Information and the UC Berkeley Center for New Media, and then at Yahoo! Research Berkeley and Yahoo! Inc., we have developed projects and prototypes which are the results of the interdisciplinary practice described above and are designed to leverage the mobile web as a sociotechnical system that is able to capture and process not only mobile media, but also mobile metadata, i.e. information about the spatial, temporal, and social context of human activity, attention, and communication in the world. We have developed and deployed several large-scale prototypes for context-aware media tagging and sharing: the Mobile Media Metadata 1 (MMM1) (*fig.2*) and MMM2 (*fig.3- 4*) prototypes at UC Berkeley in 2003 and 2004,[7] and the ZoneTag prototype (*fig.5*) at Yahoo! Research Berkeley in 2006.[8] The use of mobile media metadata has also enabled us to achieve breakthroughs in context-aware face and place recognition.[9] We have also used mobile media metadata gathered from the MMM and ZoneTag prototypes to enable us to develop new tools to visualize and interact with human photographic activity in space and time and social networks. Working with students and colleagues in the MMM2 project, we developed the Photo Level of Interest (PhotoLOI) activity visualization (*fig.6*), the PhotoCat photo browser (*fig.7*), the Assisted Metadata Propagation (AMP) semi-automatic photo annotation tool (*fig.8*), and professor Nancy Van House's Photo Elicitation Tool (*fig.9-10*).[10] At Yahoo! Research Berkeley, we developed the TagMaps visualization which also leverages the dataset of FlickrMaps (*fig.11*) in the TagMaps WorldExplorer (*fig.12*) and NightExplorer (*fig.13*).[11]

Mobile media technology connects computation to physical contexts and humans to each other, and as such provides a rich design space in which to reinvent how people, media, and technology may be understood and reconfigured in new theories, methodologies, systems, and organizations. The mediating practice in this interdisciplinary innovation process is 'design' which enables an iterative process of (de)constructing theories and (de)constructing artifacts. The result is that by understanding and constructing

MMM2 Welcome, Marc Davis

All Photos
Today (0)
Last 7 Days (0)
Last 14 Days (0)
This Month (0)
This Year (0)
All Photos (3568)

My Photos
Today (0)
Last 7 Days (0)
Last 14 Days (0)
This Month (0)
This Year (0)
All Photos (2773)

Received Photos
Today (0)
Last 7 Days (0)
Last 14 Days (0)
This Month (0)
This Year (0)
All Photos (818)

Public Photos
Today (0)
Last 7 Days (0)
Last 14 Days (0)
This Month (0)
This Year (0)
All Public Photos (165)

My Albums
2005-03-20 Oc... (51)
Lev Manovich ... (23)
CTIA 2005 Nok... (5)
110 South Hall (93)
202 South Hall (21)
Campanile in ... (15)
Mobile Monday... (24)
IS202 Assignm... (42)
Maya Soccer (21)
Yosemite Trip... (33)
BCG design me... (1)
Shoebox

Received Albums
Campanille (14)
      (Shane Ahem)
Assignment 8 (5)

All Photos : 3568 Images in this Set

Batch Actions

share	rotate:	set as:	Hide Photo Information
add to album	left / right	restricted	Square Cropped Thumbnails
delete		unrestricted	thumb size ■ ■ ■
		public	

select: all none   Thumbnails per page: 20 50 100 200 All   ⏮ ◀◀ Page 176 of 179 ▶▶ ⏭

☐ 11.04.04 12:17 PM
Marc after class

☐ 11.04.04 11:56 AM
Marc Davis in 202 (I...

☐ 11.04.04 11:32 AM
Marc Davis in 202

☐ 11.04.04 9:04 AM ✛ 👤
Bay Bridge

☐ 11.04.04 8:59 AM ✛ 👤
Clocktower

☐ 11.04.04 8:56 AM ✛ 👤
Traffic

☐ 11.04.04 8:54 AM ✛ 👤
San Francisco, CA

☐ 11.04.04 8:19 AM
Zev

☐ 11.03.04 3:57 PM 👤

☐ 11.03.04 4:49 PM 👤
Carrie

☐ 11.03.04 3:56 PM
Marc and Michael Smi...

☐ 11.02.04 12:07 PM ✛
Ray2

☐ 10.21.04 5:59 PM

☐ 10.21.04 3:04 PM
Antonia

☐ 10.21.04 1:45 PM
SIMS Studio

mobile media technologies as sociotechnical systems that connect spatially and temporally situated social beings to each other into a processing network, we make it possible to produce much better solutions to difficult technical problems such as media annotation, media sharing, face and place recognition, and behavioural targeting. This sociotechnical understanding of mobile media has coevolved with an interdisciplinary design process that not only fosters technological innovation, but also informs the reshaping of our disciplinary and methodological boundaries to make possible new theories and practices in humanities, social sciences, technology, and design that will reshape our understandings of ourselves and the role of these disciplines in shaping our future.

7 *facing page*
Screenshot detail of the PhotoCat prototype for visualizing MMM2 photos in a circular calendar filterable according to spatial, temporal, and social metadata

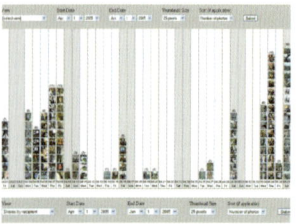

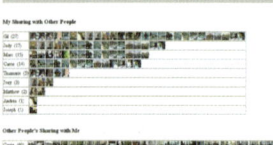

8 **Screenshot** of first version of AMP (Assisted Metadata Propagation) prototype for MMM2 photos. AMP enabled users to control the propagation of metadata from tagged photos to untagged photos according to various types of spatial, temporal, and social similarity.

9 **Screenshot from Photo Elicitation Tool** depicting a single MMM2 user's photos over time in a stacked bar graph

10 **Screenshot from Photo Elicitation Tool** depicting a single MMM2 user's photos shared to and received from other MMM2 users

Notes

[1] James Fallows, 'A Thousand Words,' *Atlantic Monthly*, April 2006, 133–135.

[2] Marc Davis and Michael Travers, 'A Brief Overview of the Narrative Intelligence Reading Group,' in *Narrative Intelligence*, eds. Michael Mateas and Phoebe Sengers (Amsterdam: John Benjamins, 2003), 27–38; Marc Davis, 'Theoretical Foundations for Experiential Systems Design,' in *Proceedings of the Eleventh Annual ACM International Conference on Multimedia* (New York: ACM Press, 2003), 45–52.

[3] Marc Davis, Simon King, Nathan Good, and Risto Sarvas. 'From Context to Content: Leveraging Context to Infer Media Metadata' in *Proceedings of the Twelfth Annual ACM International Conference on Multimedia* (New York: ACM Press, 2004), 188–195.

[4] Marc Davis 'Mobile Media Metadata: Metadata Creation System for Mobile Images (Video),' in *Video Proceedings of 12th Annual ACM International Conference on Multimedia* (New York: ACM Press, 2004); Shane Ahern, Marc Davis, Simon King, Mor Naaman, and Rahul Nair, 'Reliable, User-Contributed GSM Cell-Tower Positioning Using Context-Aware Photos,' in *Adjunct Proceedings of the Eighth International Conference on Ubiquitous Computing* (Irvine: Ubicomp, 2006).

[5] Rahul Nair, Nick Reid, and Marc Davis, 'PhotoLOI: Browsing Multi-User Photo Collections (Demonstration Description),' in *Proceedings of Thirteenth Annual ACM International Conference on Multimedia* (New York: ACM Press, 2005).

[6] Rahul Nair and Marc Davis, 'Bluetooth Pooling to Enrich Co-Presence Information,' in *Adjunct Proceedings of the Seventh International Conference on Ubiquitous Computing*, (Tokyo: Ubicomp, 2005).

[7] Regarding MMM1 see Davis et al., 'From Context to Content' and Davis, 'Mobile Media Metadata.' Regarding MMM2 see Marc Davis, John Canny, Nancy Van House, Nathan Good, Simon King, Rahul Nair, Carrie Burgener, Bruce Rinehart, Rachel Strickland, Guy Campbell, Scott Fisher, and Nick Reid, 'MMM2: Mobile Media Metadata for Media Sharing (Video),' in *Proceedings of Thirteenth Annual ACM International Conference on Multimedia* (New York, ACM Press, 2005); Shane Ahern, Simon King, and Marc Davis, 'MMM2: Mobile Media Metadata for Photo Sharing (Demonstration Description)' in *Proceedings of Thirteenth Annual ACM International Conference on Multimedia* (New York: ACM Press, 2005).

[8] See Nair and Davis, 'Bluetooth Pooling' and Shane Ahern, Marc Davis, Dean Eckles, Simon King, Mor Naaman, Rahul Nair, Mirjana Spasojevic, and Jeannie Hui-I Yan, 'ZoneTag: Designing Context-Aware Mobile Media Capture to Increase Participation,' (paper presented at the Pervasive Image Capture and Sharing: New Social

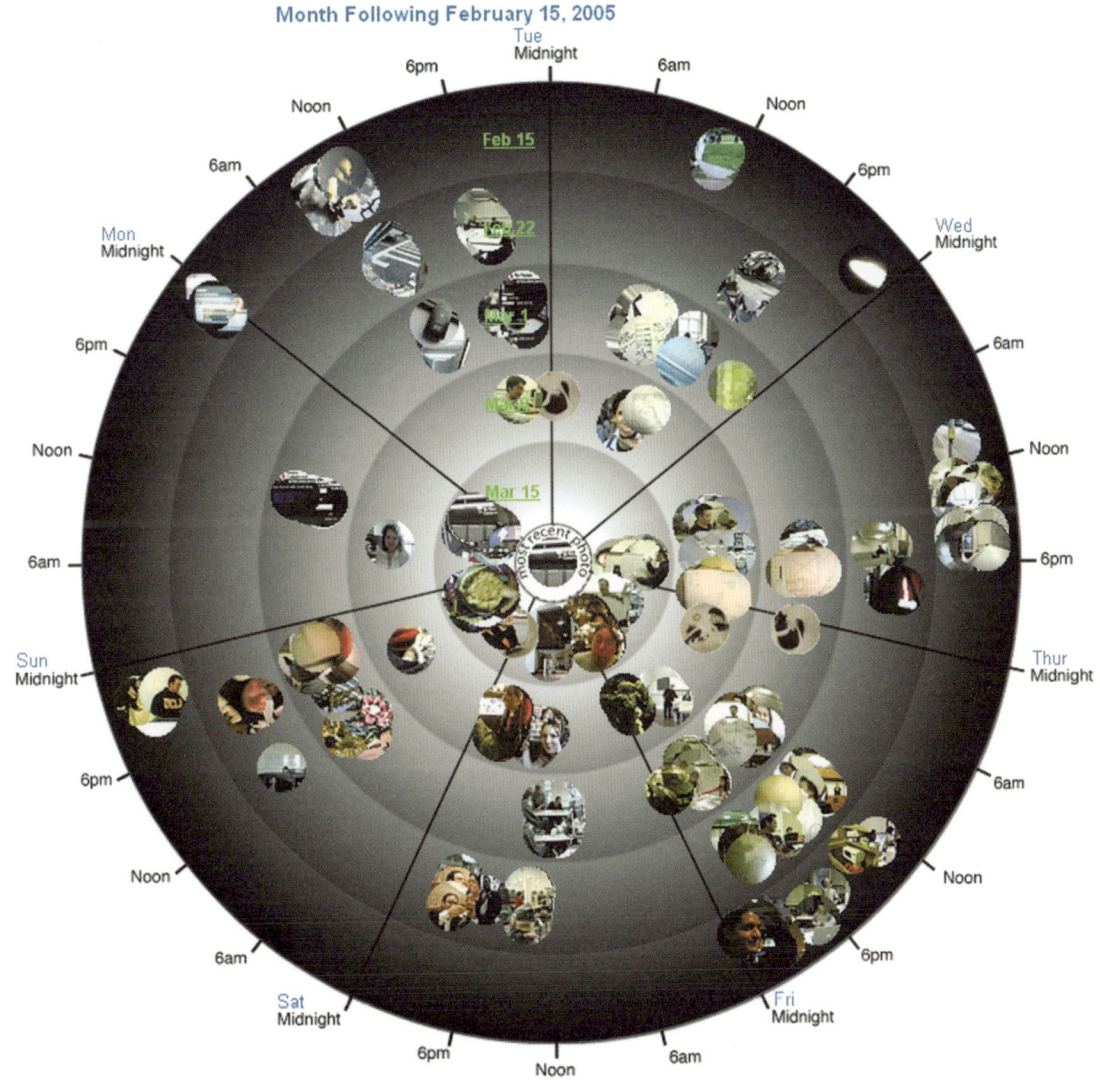

Month Following February 15, 2005

Practices and Implications for Technology Workshop (PICS 2006) at the Eighth International Conference on Ubiquitous Computing (UbiComp 2006) in Irvine, California, September 18, 2006).

[9] Marc Davis, Michael Smith, John Canny, Nathan Good, Simon King, and Rajkumar Janakiraman, 'Towards Context-Aware Face Recognition,' in *Proceedings of Thirteenth Annual ACM International Conference on Multimedia* (New York: ACM Press, 2005), 483–486; Marc Davis, Michael Smith, Fred Stentiford, Adetokunbo Bambidele, John Canny, Nathan Good, Simon King, and Rajkumar Janakiraman, 'Using Context and Similarity for Face and Location Identification,' in *Proceedings of Internet Imaging VII* (Bellingham: SPIE Press, 2006).

[10] Regarding PhotoLOI, see Nair et al., 'PhotoLOI'; regarding PhotoCat, see Carrie Burgener, Scott Fisher, Andrea Nelson, and Mike Wooldridge, 'PhotoCat,' http://www2.sims.berkeley.edu/academics/courses/is213/s05/projects/photocat/index.html; regarding the Photo Elicitation Tool see Nancy Van House, 'Interview Viz: Visualization-Assisted Photo Elicitation,' in *Extended Abstracts of the Conference on Human Factors in Computing Systems* (New York: ACM Press, 2006), 1463–1468.

[11] Alexander Jaffe, Mor Naaman, Tamir Tassa, and Marc Davis, 'Generating Summaries and Visualization for Large Collections of Geo-Referenced Photographs,' in *Proceedings of the Eighth ACM International Workshop on Multimedia Information Retrieval* (New York: ACM Press, 2006), 89–98.

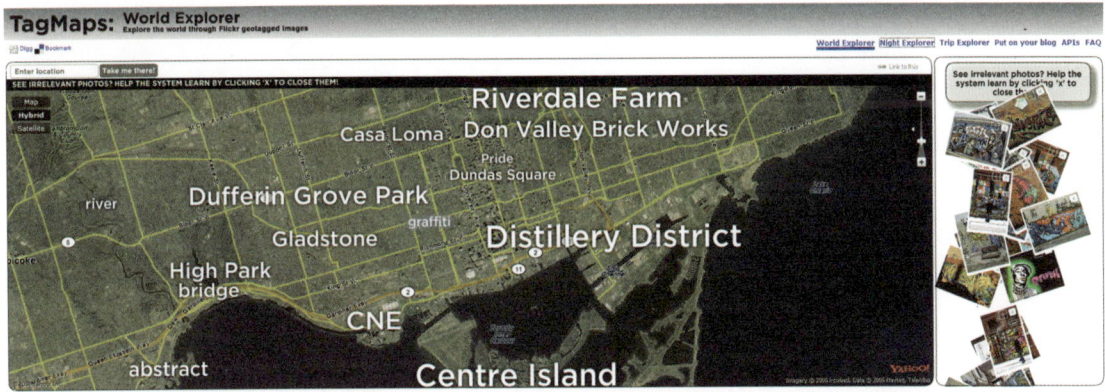

Mobile India
glimpses and opportunities

Parmesh Shahani
Mahindra and Mahindra

In this short note, I'm going to attempt an overview of the vibrant mobile space in one of the world's largest mobile markets. First I will provide a background of why India has emerged as the world's current sweetheart economy. Then I will examine some features of the Indian mobile scene.

India Shining

'India Shining' was the name of the incumbent right-wing BJP (Bharatiya Janata Party or 'Indian People's Party') re-election campaign in the 2004 general elections. They unexpectedly lost but their campaign's title continues to resonate, and today it seems that everywhere you look (at least in English-speaking India and abroad) India is the flavour of the season.[1] There is a steady flow of country presidents and corporate CEOs through Bombay, Bangalore, and New Delhi. Symbolically, at the 2006 World Economic Forum at Davos, Switzerland, India's business leaders, politicians, and Bollywood stars combined efforts to brand 'India Everywhere,' a blitzkrieg that had, among others, summit chair Klaus Schwab dancing in a turban and shawl and extolling India's virtues.[2] A confluence of three forces has fashioned this 'perfect storm.'

Appealing demographics

In 1985, 93 percent of India's population lived on less than one dollar per day. The number is projected to decline to 22 percent by 2025.[3] More importantly, as recent research from the McKinsey Global Institute indicates, within the next two decades, there will be an Indian middle class of over 550 million people.[4] Within this same period, approximately 120 million Indians will begin to have discretionary income. India is especially attractive because it has one of the youngest populations in the world and can reap a potential 'demographic dividend' from this young population—in fact, by 2020, the average Indian will be only 29 years of age, versus 37 for the average Chinese, and 45 and 48 years respectively for the average European and Japanese.[5]

facing page
1 **Cars passing by at night**

Booming economy

India's GDP has been growing impressively at 9 percent per annum in the recent past. From being shunned by investors, it has morphed into a desirable global market—A. T. Kearney's 2005 Foreign Direct Investment Index ranked it as the second most attractive country in the world to invest in. (China was first.)[6] India's higher political profile and nuclear power status have added weight to its economic stature. While trends like the millennial contract labour and outsourcing boom made India a buzzword in the West, what's really appealing to investors today is not India as a source of cheap labour, but India as a market and India as a source of global innovation. Currency convertibility, the boom in retail finance, and the imminent growth of power and infrastructure are all changing the country's complexion as a marketplace. Take the retail sector, for instance, that is growing at an estimated $27 billion a year with most international players like Wal-mart and Carrefour in the picture, ready to battle local heavyweights like Reliance and Big Bazaar.[7] India is also emerging as the go-to place for global best-practice solutions. For example, a financial services company called ICICI has launched a diabetes management policy where the insured patient's premium gets reduced if the disease is managed well—the first of its kind anywhere in the world.[8]

Soft power

'Soft power' is a term coined by Harvard professor Joseph Nye to refer to a country's cultural influence over others.[9] India's soft power (exemplified by Bollywood, Indian food, Bhangra music, cricket, and so on) has been well utilized by its government and corporations to add value to the feel-good India story. Pervasive 24-7 media and the presence of a strong Indian diaspora have ensured that India-related stories resonate globally—in the homeland as well as in adopted lands. There was controversy on the UK TV show *Celebrity Big Brother* in early 2007, when presumably racist comments by a fellow housemate towards Indian actress Shilpa Shetty circulated all over the world driven by global news networks and a jingoistic, vocal diaspora. Online and offline protests, blog posts, and televised anger capsules from India and the UK made everyone from the Indian and British prime ministers to the British monarch sit up and take notice. Eventually, Shetty won the show finale by a public vote of 63 percent.[10]

Mobile Nation

Now that we've set the scene, welcome to mobile India!

Prior to 1992, the government of India had a monopoly over telecommunications, and there were only about five million fixed-line telephones in India in 1990.[11] As part of the economic reforms process, the telecommunications sector was liberalized in 1992 and private sector participation was encouraged, especially in the mobile services sector. However, by 1997, despite the attractiveness of the Indian market, there were only about 800,000 cellular subscribers in the country. The government then

"You are either a customer or a product."

Rahul Bose

Ctrl + Alt + Del

2 *Ctrl+Alt+Del* India's first made-for-mobile film

revamped the National Telecom Policy, which translated into lower tariffs for consumers, greater subscriber uptake for operators, and increased cellular coverage in the country. These and the subsequent progressive government moves (allowing CDMA cellular coverage and enforcing the calling party pays (CPP) protocol, effectively making all incoming calls free) saw the cellular subscriber base increase to approximately 5.5 million users by the end of 2001, which is when the party really began to rock.[12] The number of cellphone subscribers sharply rose to reach 50.8 million in February 2005 and over 120 million by August 2006,[13] by which time India had become the fastest-growing cellular market in the world.[14] The figure is expected to surpass 500 million by 2010.[15] (Comparatively, there were only approximately 30 million Internet users as of 2006.)[16] Attracted by these vast numbers, the world's mobile biggies have begun flowing to India. In March 2007, Vodafone purchased 67 percent of service provider Hutchison Essar for $11.1 billion.[17] Others like DoCoMo are also eyeing the market for an opportunity to enter.[18] At the same time, Indian telecom companies are expanding globally into other Asian and African markets.

Market

Features of India's mobile market:

+ Ultra-low-cost handsets dominate (priced below $20).
+ 80 percent of the market consists of prepaid connections.
+ Voice still dominates, but data- and voice-based value-added services are rising fast.
+ Telecom and broadband is growing simultaneously in rural areas (many rural citizens will experience Internet first on a mobile).
+ Indian-language mobile content increasing.

MIT professor Eric Von Hippel has famously written about lead users or early adopters of products or technologies also being early adapters.[19] We can see several cases of this adaptation in mobile India. For example, the famous Siddhivinayak Temple in Bombay has a live *aarti* (a ceremonial ritual based on the lighting of oil lamps) webcast, online booking of prayer rituals, and *prasad* (sweets consumed by devotees after first being offered to the deity) delivery both within India and abroad (via FedEx or other courier services). There is also a service to process donations and *prasad* requests via text messaging. The temple has partnerships with most of the major cellphone companies in the country for text alerts of prayers and *aartis*, downloads of deity wallpapers, ring tones, logos, and ecards.

It is pertinent to remember that 70 percent of India continues to be rural and agrarian. The country has twenty-four languages, 1,642 dialects, all major religions, and 20,000 ethnic groups. There are 638,691 villages, 5,164 towns, and 138,000 post offices in rural India, and 25 million TV households.[20] Historically, this rural market has been underserved.

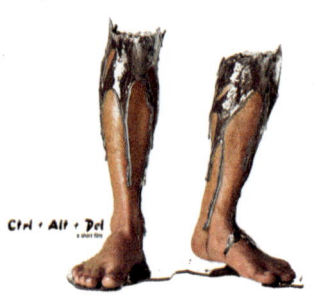

3 *Ctrl+Alt+Del* India's first made-for-mobile film

University of Michigan professor C. K. Prahalad has demonstrated that there is a 'fortune at the bottom of the pyramid'—poor people can be lucrative consumers if a company can market its products effectively to them.[21] Small-town, rural, and poor India is an immense market. The indications are strong. For example, 55 percent of the life insurance policies sold by the Life Insurance Corporation of India have been sold to rural India. Forty-one million Kisan (farmer) credit cards have been issued to Indian farmers.[22]

Sachet marketing (making branded goods accessible to consumers with limited cash by packaging them in small, inexpensive quantities) works in India—whether with shampoos or with mobile services. Selling tiny, one-off amounts for low prices has been tremendously successful for talk-time recharges as well as value-added services like ring tones. Arvind Rao, CEO of the mobile value-added services company On Mobile, notes in a study that three out of five callers for a television dial-in show are from towns with a population of less than 500,000 people, or villages.[23] Star 7827—a text-message IVR service from Rupert Murdoch's Star network—finds that it garners a 70 percent audience share from India's smaller towns and villages for its audio mobisodes of popular daily TV soaps.[24] India is also fast emerging as a forerunner in using the cellphone as a tool to access the Internet, with the country home to the fourth-largest population browsing the World Wide Web through their mobile handsets. One in every eleven people logging onto the web through mobiles across the world is Indian.[25]

Opportunities

Hundreds of interesting companies are springing up in mobile India. These include cellular service providers (Vodafone Essar, Airtel, Reliance), telecom service companies (Tech Mahindra), retail content producers and aggregators (Hungama Mobile), mobile payment enablers (Paymate, Obopay, MChek), mobile communities (Tadkalive), literacy agents (Tinfo Mobile), job portals (Clickjobs)… one could go on and on. All this activity naturally makes India a very compelling place to do research. There is enough quantitative work being done, largely about usage habits. Surveys like the 2007 *Voice and Data* magazine roundup of 4,524 urban mobile users indicates that that 43 percent of those surveyed like to change their phone each year, and 58.4 percent download ring tones.[26] Useful stuff, surely, but what is sorely missing, or at least, very scarce, is good qualitative research, of the kind carried out by doctoral candidate Carolyn Wei (University of Washington) in Bangalore on mobile phones and courtship behaviour between lovers, married couples, and arranged marriage candidates.[27]

facing page
4 **Scenes from** *Ctrl+Alt+Del*

Notes

[1] 'BJP Admits "India Shining" Error,' *BBC News*, May 28, 2004, http://news.bbc.co.uk/2/hi/south_asia/3756387.stm; 'A campaign that lost sheen,' *The Hindu*, October 3, 2004, www.hindu.com/mag/2004/10/03/stories/2004100300160200.htm.

[2] Fareed Zakaria, 'India Rising,' *Newsweek*, March 6, 2006, www.msnbc.msn.com/id/11571348/site/newsweek/.

[3] 'The Second Asia Shock,' *Newsweek*, May 28, 2007, international edition, www.msnbc.msn.com/id/18753951/site/newsweek/.

[4] Diana Farrell and Eric Beinhocker, 'The World's Next Big Spenders: How India's Rising and Unique Middle Class will Reshape Global Consumer Markets,' *Newsweek*, May 19, 2007, international edition, www.msnbc.msn.com/id/18753700/site/newsweek/page/0/.

[5] C. P. Chandrasekhar and Jayati Ghosh, 'India's Potential "Demographic Dividend,"' *Hindu Business Line*, January 17, 2006, www.blonnet.com/2006/01/17/stories/2006011701531100.htm.

[6] A. T. Kearney FDI Confidence Index 2005, www.atkearney.com/main.taf?p=5,3,1,140,1

[7] Emily Wax, 'In India, a Retail Revolution Takes Hold; Small Vendors Feel Squeeze of Chains' *Washington Post*, May 23, 2007, www.washingtonpost.com/wp-dyn/content/article/2007/05/22/AR2007052201409.html.

[8] See details of the ICICI Diabetes Care policy at ICICI Prudential, 'Why Diabetes Care?' www.iciciprulife.com/public/Health-plans/DiabCare.htm.

[9] See Joseph S. Nye Jr., *Soft Power: The Means to Success in World Politics* (New York: Public Affairs, 2004).

[10] See 'Politicians enter Big Brother Row,' *BBC News Online*, January 17, 2007, http://news.bbc.co.uk/2/hi/entertainment/6269953.stm; 'Shetty wins Celebrity Big Brother,' *BBC News Online*, January 29, 2007, http://news.bbc.co.uk/2/hi/entertainment/6308443.stm; Christopher Morgan, 'Big Brother's Shilpa gives Queen Talk on Tolerance' *Sunday Times*, March 11, 2007, www.timesonline.oo.uk/tol/news/uk/article1496825.ece.

[11] Gurcharan Das, *India Unbound* (New Delhi: Penguin/Viking, 2000), 9.

[12] Information and statistics sourced from the official website of the Cellular Operators Association of India (COAI), an industry association comprising all the cellular phone companies operating in India. See http://coai.com/.

[13] In October 2004, the number of mobile phones in the country surpassed the number of landline users (44 million) for the first time. See Arindam Mukherjee, '98 Tra La La 1000,' *Outlook*, April 4, 2005, www.outlookindia.com/full.asp?fodname=20050404&fname=VTelecom+%28F%29&sid=1; Anand Parthasarthy, 'Mobile Phone Growth Signals India's Telecom Maturity,' *The Hindu*, October 16, 2004, www.hindu.com/2004/10/16/stories/2004101603401300.htm.

[14] See Saritha Rai, 'India Leads World in Cellphone Expansion' New York *Times*, September 15, 2006, www.iht.com/articles/2006/09/15/business/cell.php.

15 See Indrajit Basu, 'India's New Telecom Callers,' *Washington Times*, June 25, 2004, http://washingtontimes.com/upi-breaking/20040624-010347-8465r.htm.

16 Statistics sourced from Juxt Consult, 'India Online Survey 2007,' http://juxtconsult.com/syndicated_research/indiaonline2007/ internet_report_main.asp.

17 Randeep Ramesh and Sanjay Jha, 'Vodafone Signs Deal for Control of India's Largest Mobile Firm,' *Guardian*, March 16, 2007, http://business.guardian.co.uk/story/0,,2035267,00.html.

18 Nivedita Mookerji, 'DoCoMo May Come Calling,' *DNA India*, June 19, 2007, www.dnaindia.com/report.asp?NewsID=1104237.

19 See Eric von Hippel, *Democratizing Innovation* (Cambridge: MIT Press, 2005).

20 Statistics sources: FICCI Rural Marketing Summit 2005, cited in Rajiv Hiranadani, 'Rural India: The Diversity Within,' (presentation, Internet and Mobile Association of India Digital Summit, New Delhi, January 18–19, 2007) and Virendra Gupta, 'Driving Adoption Beyond Metros,' (presentation, Internet and Mobile Association of India Digital Summit, January 18–19, 2007). Both presentations accessible at http://www.contentsutra.com.

21 See C. K. Prahalad, *The Fortune at the Bottom of the Pyramid: Eradicating Poverty Through Profits* (London: Wharton School Publishing, 2004).

22 M. V. Nair, 'Indian Banking: Shaping an Economic Powerhouse,' (presentation, Federation of Indian Chambers of Commerce and Industry, New Delhi, July 17, 2006) archived at www.ficci.com/media-room/speeches-presentations/2006/ jul/kolkata/Nair.ppt.

23 Arvind Rao, 'On Mobile,' (presentation, Internet and Mobile Association of India Digital Summit, January 18–19, 2007), archived at www.contentsutra.com.

24 'Star 7827 Voice platform Crosses 2 Million Minutes in 2 Months,' *Moneycontrol.com*, November 9, 2006, www.moneycontrol.com/india/news/pressnews/ star7827voice/star7827voiceplatformcrosses2mnminutes2mths/ market/stocks/article/250267.

25 'India sees two-fold jump in mobile web usage: Study,' *DNA India*, June 6, 2007, www.dnaindia.com/report.asp?NewsID=1101660.

26 'India Connects,' *Hindustan Times*, January 16, 2007.

27 See Hannah Hickey, 'Mobile Phones Facilitate Romance in Modern India,' UNWNews.org, February 12, 2007, http://uwnews.washington.edu/ni/article.asp?articleID=30477.

(Im)Mobile Nation
the iterative design process

Maroussia Lévesque and Jason Lewis
Concordia University

1 *Cityspeak* interactive text installation. Foreground: Web interface on a personal digital assistant. Background: *Cityspeak* projection. Oboro gallery, Montreal, 2007

Cityspeak and *Citywide* are sibling projects that explore *situated wireless*: mobile networked technologies used in relatively immobile contexts to augment conversations among people located in the same space. *Cityspeak* uses large-scale public displays to host text messages sent by participants' mobile phones and wireless PDAs, using prompts and subtle provocations to encourage participants to talk to one another about the particular context in which they find themselves. *Citywide* moves the locus of interaction to the personal space of the laptop screen, where participants enter a chat environment focused on a particular hotspot or group of hotspots. We have created over a dozen *Cityspeak* installations in contexts ranging from workshops to festivals to nightclubs, and *Citywide* has been active across Montreal's free wireless network since the spring of 2007. Here we review our experience with these two projects, discuss how they are effective at promoting conversations about a particular space, and contemplate how they may evolve in the future.

What is radically new about mobile technologies? What new possibilities do they open for (re)engaging with others and ourselves? These questions are what launched the *Cityspeak* and *Citywide* projects at Obx Laboratory for Experimental Media.[1] For the past three years, we've used these questions to guide our intuitions and suspicions about the potential within mobile technologies for new modes of social interaction. In the process of developing the projects we developed the notion of situated wireless, which is where these technologies are used to reinforce localities via site-specific interventions rather than as nomadic nodes on a dispersed, decontextualized network.

Cellphones and laptops enable anytime, anywhere access to data from around the globe, but they are equally useful for deeply exploring the specifics of a particular location. How can mobile input methods be connected to public displays for site-specific explorations? *Cityspeak* and *Citywide* are interactive text projects aimed at altering and augmenting locations where the global becomes local.

Cityspeak

facing page

2 *Cityspeak* large-scale projection. Concordia University, Montreal, 2006

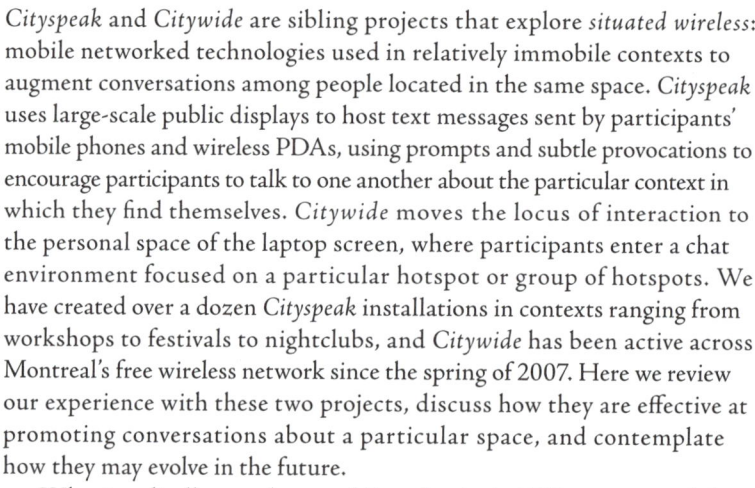

Cityspeak is an interactive text installation that can be accessed by sending an SMS message with a cellphone, or inputting text through a web form with

nd sources of m
spot or Tuvalu? th
heek hello find the
come a bac
s underline
messages. Vive
ician 514-991-612
rdire! je bois une
ple have pda's or

any web-ready device. On the display itself, messages first float in the foreground, then get pulled into a chaotic mass at the bottom right before being integrated into a history of old messages scrolling in the background. The history is constantly updated as new messages are sent, providing contextual information that new senders can reinforce, denounce, or ignore.

Screens: Facade 2.0

> We live in societies which are enveloped in and saturated by the media: most importantly, it is difficult to escape the influence of the screen which now stares at us from so many mundane locations—from almost every room in the house to doctor's waiting rooms, from airport waiting lounges to shops and shopping malls, from bars to many workplaces.[2]

Nigel Thrift proposes that the capacity of real-time media to affect urban inhabitants is a product of the inability of the viewer to choose *not* to attend to the message being broadcast. Moving screens are an integral part of a commercial content continuum in which we bathe everyday. The result is that we suffer the media rather than consume it.

Cityspeak is designed to be displayed on public screens in order to take advantage of this growing omnipresence and to explore the still-emerging affective qualities of these screens. From Greek monuments to medieval cathedrals to modernist skyscrapers, spatial (infra)structures are created in order to touch the viewer emotionally, inspiring fear, respect, or admiration. Up until recently, the facades of such constructs have been fairly static, their content hard-coded into stone, concrete, and steel. With the advent of screens on top, beside, or completely replacing the facade, we are seeing more ephemeral messages that transmit the soup of the day, the hot *flavas* of the co-opted underground, and the fleeting desires of fashion. This ever-changing face of the postmodern facade threatens to relegate it to a vacuous, Baudrillardian hyperreality.[3] Public screens are 'Facades 2.0,' constantly updating themselves with broadcasts that alter our perception of time and place, suggesting our wardrobe is out of date, our computer equipment obsolete, our cars so last year.

Yet we are not Luddites: we want to bend screens to other ends, not dispose of them. *Cityspeak* intervenes to provide a channel for talking back to these urban screens, to shout up against the sensory overload and counter the passivity of simply putting up with the screens shouting down at us.

Threat of Text

While the number of large-scale public screens is growing, the percentage supporting non-commercial use remains quite small.[4] Most big-screen content is pre-programmed, so of that small percentage there is an even smaller number that invite the audience to interact in real time. Moreover, the few real-time interactive works that do exist are seldom text-based. As we pursued public display options for *Cityspeak*, we realized that the literal nature of text is perceived as a threat.

3 *Cityspeak* public installation. Soukmachines interdisciplinary art festival, Montreal, 2006

Many venues have expressed interest in the idea of *Cityspeak*, but upon learning that it traffics in real-time text messages venue operators often became resistant. Their resistance is due to concerns about slander, obscene language, and the fact that work such as this—which empowers end-users to deploy their personal mobile devices to inscribe urban surfaces and enact an agenda different from the standard commercial one—is perceived as risky. We interpret this resistance as informal confirmation of the perceived power of employing text in this manner.

Future Steps

Perhaps public space is a fiction, the spatial projection of a democratic ideal yet to materialize. Our final *Cityspeak* installation will test this notion. It will be at Victory Park, in Dallas, Texas.[5] This new residential hotel, shopping, and sports complex, featuring a facade with eight large-scale (15 feet by 26 feet) moving screens, epitomizes the contemporary tension between public and private: the plaza in the middle of the complex is city property, but it is managed by the complex developers. Part of our interest in mounting *Cityspeak* at Victory Park is precisely because it is not clear whether this plaza is a public space—defined simply as an inclusive, heterogeneous, and uncensored space for free interaction[6]—or a private consuming environment which *emulates* public space but comes complete with pervasive camera surveillance, strict regulations of facade appearance and private security responsible for the removal of the unwanted or unsavory. We plan on using *Cityspeak* to push on the question of how such spaces are perceived, used, and managed.

Citywide

Citywide is the result of our work on *Cityspeak* leading us to focus on the interaction between mobile technologies and a particular place.[8] Through our participation in the Mobile Digital Commons Network (MDCN) we were introduced to Île Sans Fil (ISF), a free wireless collective in Montreal.[9] We were engaged by their vision of wireless access points as community nodes around which various sorts of grassroots activities could be organized and presented, and by a use profile that suggests interesting congregations of people. We wondered how we could reach these people with a version of *Cityspeak*'s shared space for public commentary that would be appropriate to the character of the various ISF hotspot locations.

Citywide is the product of that curiosity. *Citywide* is a local chat interface for each ISF hotspot, designed as an additional layer of communication to support the micro-communities that form around the cafés, bars, and bookstores that host nodes on the free wireless network. It is the specificities of each location that struck us as more lush terrain for mobile communities than the abstractions of a global village where anyone anywhere can be anything.

How it Works

We designed *Citywide* specifically for people using laptops. The application interface is written in Flash to allow it to be run on any browser, and the back end connects together people accessing the net through the same wireless access point.

Citywide works like any chatroom. One can send a message, select a thread to respond to, and view previously written messages. However, access to the chatroom is restricted to users logged-in via a given hotspot. In other words, only the people at Café Laïka can enter the Laïka *Citywide* chatroom, and only the people at Café Utopik can enter the Utopik chatroom.[10]

Visitors can access *Citywide* via the login page that ISF uses to authenticate users onto their network. Each ISF hotspot has a customized homepage, including a shoutbox. When a user sends a message via the shoutbox she gets taken to the *Citywide* chat space, where she exchanges messages with other users at the same hotspot. The *Citywide* interface allows her to personalize the appearance of her text messages, and to get an overview of the number of messages sent from all users and the total number of users for the past twelve hours.

4 ***Citywide* chat interface**
Laïka café, Montreal, 2007

The Future of Situated Wireless

At the time of publication, *Citywide* has just been launched over all 106 Île Sans Fil hotspots. This initial deployment will serve as a test of the capacity of situated wireless to forge or reinforce interaction between loosely coupled co-located users, as well as provide data to compare with our experience with *Cityspeak*. In both works, users are anonymous to other users (unless they choose to declare their real identity through message content, or, in the case of *Citywide*, user name.) Yet their messages are public in different ways, with *Cityspeak* providing a display space that is public to everybody regardless of whether they are participating, and *Citywide* providing a display space shared only by the users.

Both *Cityspeak* and *Citywide* use physical location as a starting point, as an asset to reinforce, a fertile space of potentialities. In these ways *Cityspeak* and *Citywide* invert the anywhere/anytime promises of wireless mobile technology hinted at by McLuhan and commoditized by commercial carriers. If mobile devices are used to access the same data from anywhere, they come to act as blinders or filters suppressing the difference between places. Our experience with these projects leads us to believe that mobile technologies can be used to re-embody and resituate space, not to dislocate or erase it. If mobile tech can be used to reassert the specificity of place and those present in it, they would function as a lens, a flashlight, a means of augmenting instead of replacing. This is the rich terrain of mobile art we are interested in exploring further.

Acknowledgements

The authors thank Yannick Assogba, Lucie Bélanger, Lysanne Bellemare, David Bouchard, Hugues Bruyère, Zehuan Liu, Alexander Taler, and Elie Zananiri for their tremendous help in coding, shaping, and thinking through these projects. We would also like to thank Canadian Heritage for their generous financial support.

Notes

[1] For more information see www.obxlabs.net.

[2] Nigel Thrift, 'Spatial Politics of Affect,' *Geografiska Annaler, Series B* 86 (2004): 57–78.

[3] Jean Baudrillard, *Simulacra and Simulation* (Ann Arbor: University of Michigan Press, 1994).

[4] Creative Time, 'The 59th Minute: Video Art on the NBC Astrovision by Panasonic,' www.creativetime.org/programs/archive/59; Realities:United, 'Bix Communicative Display Skin for the Kunsthaus Graz, Austria,' www.bix.at.

[5] For more information on the development see www.victorypark.com.

[6] Pieter Boeder, 'Habermas' Heritage: The Future of the Public Sphere in the Network Society,' *First Monday* 10 no. 9 (2005), http://firstmonday.org/issues/issue10_9/boeder/index.html.

[7] Stephen Graham and Simon Marvin, *Splintering Urbanism: Networked Infrastructures, Technological Mobilities and the Urban Condition* (New York: Routledge, 2001).

[8] For more information on *Citywide* see www.cwide.org.

[9] For more information on the MDCN see www.mobilelab.ca/mdcn/. An overview of the wireless network is available at www.ilesansfil.org.

[10] For more information on these venues see www.laikamontreal.com and http://lutopik.org.

[11] 'Citywide Showreel,' www.cwide.org/goodies/video/Citywide.mov.

Research and Design for Mobile Platforms
a walk in the park

Martha Ladly
Ontario College of Art & Design

How can mobile technology enhance an urban or wilderness park experience? This is the primary research question that drives the projects designed within the Mobile Digital Commons Network (MDCN).[1] For the *Park Walk* project, MDCN members were also interested to find out how we could tell stories and convey specifically located information in a park with the use of mobile technology. Could participants actively engage in the authorship and creation of these stories and informational content? And how would we design a self-guided experience so that visitors using mobile technology could discover and participate in activities located in the park? The *Park Walk* project continues to be premised on the social norm that park visitors must leave no trace, and take nothing away from the park setting. Beyond socially responsible behaviour, this premise is legally enforced in wilderness areas, nature preserves, and national parks such as Banff National Park, an important research site for the project. The convention carries through to the urban park, and operates in locales such as High Park and Mount Royal Park,[2] where there have been concerted attempts to preserve pockets of wilderness in the urban settings of Toronto and Montreal, respectively. These sites proved extremely useful as local test beds, standing in as miniature urban models of Canada's great wilderness parks. A third site, the urban estate of Grange Park in Toronto,[3] came into its own as a location for designing and testing mobile interventions in an inner-city park.

In the preliminary research phase conducted in Banff National Park, High Park, and Mount Royal Park, we verified through interviews our belief that many park visitors would like to break the established social norm; they intuitively seek to collect artifacts or souvenirs of their experience of the park. Many said they would also enjoy the ability to leave a virtual 'signature' such as a photo or audio clip that they could revisit and share with friends and other park visitors. In High Park, we conducted a small survey of visitors (*fig. 4*) and found that 100 percent had brought some form of mobile technology with them to the park (*fig. 5*). This was a surprising result; the sample consisted of 60 percent females with an average age of 37. The technology they brought with them ranged from mobile phones to cameras and gaming devices (*fig. 6*), encouraging us to

[1] **The MDCN** connects research, arts, and industry focused on mobile, wireless, digital technologies in Canada

[2] **High Park,** the former home of architect, surveyor, and engineer John Howard, is a tract of 398 acres of ancient oak savannah, woodland, fields, and gardens in the west end of Toronto ceded to the city for the enjoyment of its citizens in 1873; **Mount Royal Park** is located at Mount Royal, the highest point in Montreal, and includes one of the city's largest green spaces. The park is mostly native woodland, fields, and trails laid out and designed by Frederick Law Olmsted, the designer of New York City's Central Park. It was inaugurated in 1876.

[3] **Grange Park** was formerly the estate of Darcy Bolton Jr. His home, the Grange, still stands adjacent to the Art Gallery of Ontario and the Ontario College of Art & Design.

facing page
1 *Park Walk* **Project**
Spring Creek Trail Map and activities menus on mobile devices

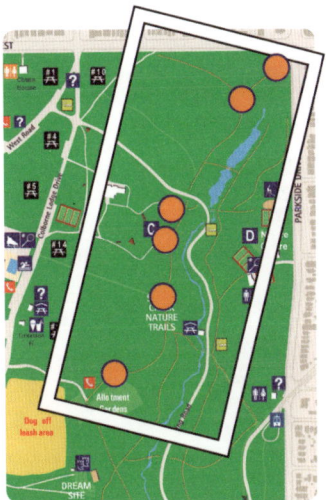

Toronto High Park

surmise that most park visitors carried some form of mobile technology, and were comfortable using it in the park. In order to find out where people were playing, walking, biking, and recreating in the park, we asked them to identify sectors they had visited that day (*fig. 7*), and we used this information to identify potential locations for our experience design.

Mobile Technologies in the Park

Early on in the research we made a decision to employ GPS devices and mobile telephones as the technologies that would enhance the *Park Walk* experience.[4] *Park Walk* employs GPS coordinates mapped on various trails in Toronto's Grange Park and High Park, and on the Hoodoos Trail in Banff National Park. Participants use a specially programmed application on their mobile phone, connected via Bluetooth with a GPS device,[5] to see images, hear stories, and find out more about the park and the trail they are walking. Geographically specific images of maps, graphics, photographs, and animations are mapped to locations in the park, and audiovisual narratives are triggered as the visitor enters the concentric circles of GPS hotspot locations that have been mapped to the site. The images and stories they see and hear on their mobile phones pertain directly to the historical, cultural, and natural characteristics of that location. An additional feature is a pictorial identification system, installed on the mobile, to the local flora and fauna. A far cry from a didactic 'field guide,' the application lays a veil of relevant information over a geographic area, rooting it firmly to specific locations of interest. The availability of location-specific information and stories is notified to the visitor by the GPS signal triggering the vibration or ring of their mobile phone when they pass through the pertinent coordinates. In this way, visitors are able to pluck information about their location as if from the air, if they so choose.

Park Walk uses a handheld GPS device, linked via Bluetooth with built-in recording features of an iMate mobile phone, to enable our important research finding regarding the leaving of traces and the capturing of souvenirs. This innovation, currently in development, allows participants to capture their own experiences in the park (via specific GPS-located data), which are stored locally on the phone and later uploaded to the MDCN server. This facilitates an editorial review by the project team, the interweaving of this new content into the application, and consequently the ability for visitors to experience the new data when they return to or visit that location for the first time. Visitors may capture personal data in the location, and retrieve it later as a souvenir or *aide-mémoire* and a recording that may be a new layer in the virtual space (their personal signature on the tree, the view in changing seasons, or the documentation of physical changes to the park, such as the erection of the new Frank Gehry addition to the Art Gallery of Ontario, overlooking Grange Park) for other visitors to experience and enjoy. In this way, the space is visited and refreshed with digital artifacts that are woven back into the virtual space, offering the potential of psychogeographical layers of a place that may be created over time.

2-3 **Mobile technology and GPS**

4 **GPS** is a satellite-based navigation system made up of a network of twenty-four satellites placed into orbit by the US Department of Defense. GPS was originally intended for military applications, but in the 1980s the government made the system available for civilian use. Garmin, 'About GPS,' www.garmin.com/aboutGPS/.

5 **Bluetooth** wireless technology is a short-range communications system intended to replace the cables connecting portable and/or fixed electronic devices. Bluetooth, 'How Bluetooth Technology Works,' www.bluetooth.com/Bluetooth/ Learn/Works/.

facing page
4-7 ***Park Walk* survey 26**
20 Participants, High Park, Toronto, June 2006, visualised by Bryn Reed-Ludlow

Park Walk Survey, Chart 1
Questions: How old are you?

Nō	M/F	Age
1	M	69
2	F	23
3	F	45
4	F	25
5	F	35
6	F	50
7	F	30
8	F	30
9	F	35
10	F	20
11	M	35
12	M	45
13	M	30
14	F	20
15	F	33
16	M	05
17	M	13
18	M	35
19	M	65
20	M	61

60% females	36.7 Average age
40% males	of participant

Park Walk Survey, Chart 4
Ratio of Women to Men Using Technology in High Park

Tech Device	Women	Men
Cellphone	60%	40%
Digital Camera	70%	20%
Gaming Device	10%	0%

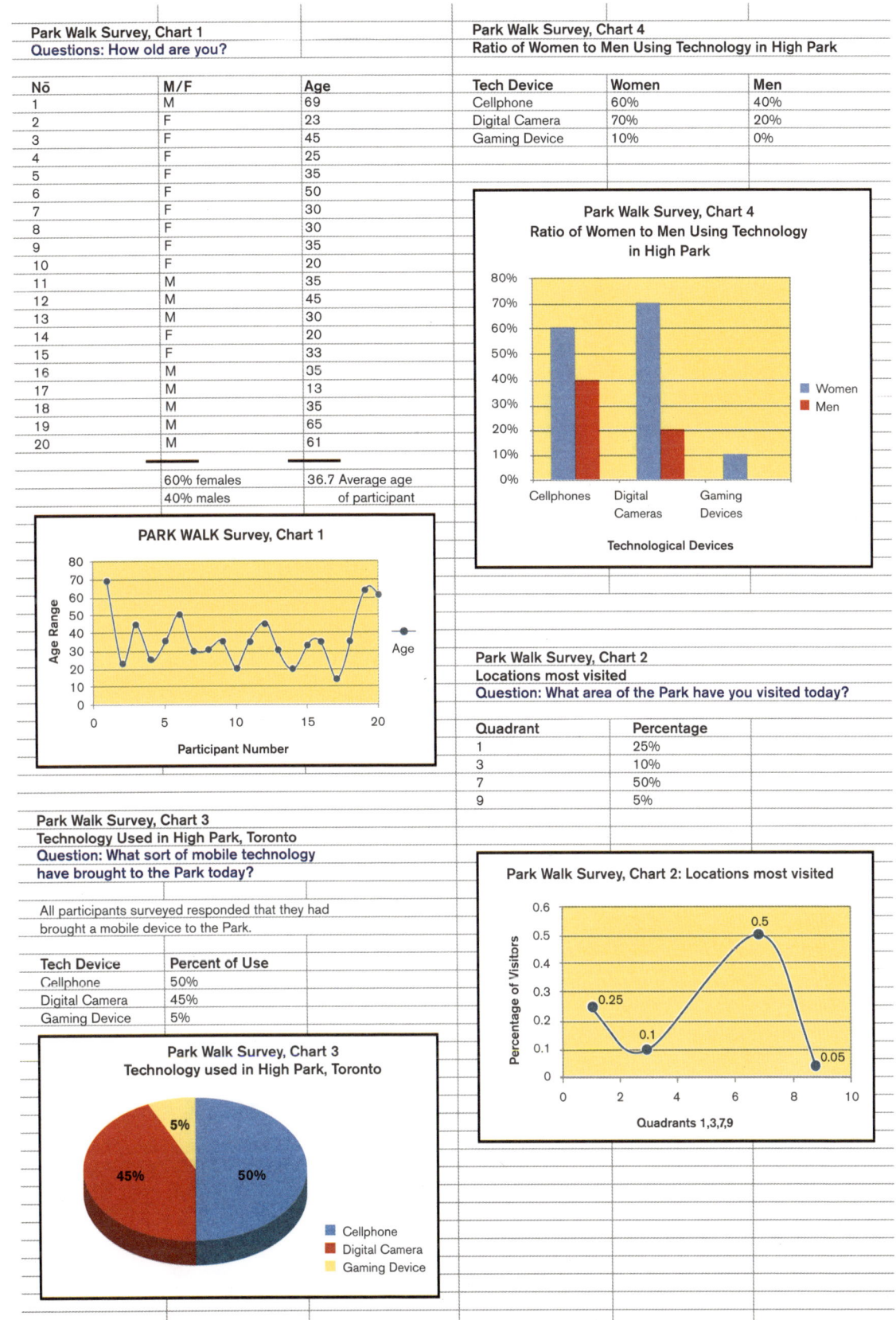

Park Walk Survey, Chart 2
Locations most visited
Question: What area of the Park have you visited today?

Quadrant	Percentage
1	25%
3	10%
7	50%
9	5%

Park Walk Survey, Chart 3
Technology Used in High Park, Toronto
Question: What sort of mobile technology have brought to the Park today?

All participants surveyed responded that they had brought a mobile device to the Park.

Tech Device	Percent of Use
Cellphone	50%
Digital Camera	45%
Gaming Device	5%

Secondary questions arise once one decides to design for the capture and implementation of user-generated content. How should this new content be effectively structured, and what levels of editorial intervention are needed? Is it possible to facilitate personalized experiences? What level of expectation does this raise in the user? How much of this sort of 'work' does the visitor wish to participate in? These are also questions of design intention, and as the designer I wish to maintain control of the experience at an artistic level, to integrate new material harmoniously; so a level of editorial authority is required.

Mobile Design Challenges

When I discuss locative media, I am referring to technologically mediated forms of communication and recreation which reference real-world locations, thus triggering social interactions.[6] The term ties together critical perspectives, research questions, and practices that allow for the annotation and navigation of the environment. This has opened up new discussions about the relationship between consciousness and place, between people and place, and between groups of people in located situations. Design for multi-modal and multi-dimensional experience challenges current design methodologies.[7] Interactions in these environments are not reducible to discrete tasks; rather they involve fluid activities on multiple devices and at multiple social and operational levels in which goals and purposes overlap.[8] The social nature of mobile use requires new and innovative research and design methodologies such as participatory design and improvisation.

Participatory design introduces a variety of techniques and allows for a modulation of levels of participation in the design process. MDCN researchers have devised a variety of techniques to facilitate the communication and testing of new technological possibilities with users. These techniques include the use of mock-ups and role-playing activities, as well as photo, video, and animation.[9] We have innovated with paper prototypes and semi-functional or fully functioning demonstrations to simulate the patterns of interaction with a new interface. This helps us to project outcomes, imagine alternatives, and refine iterations of a design. During the design phase, recommendations from user-testers on the design team are incorporated into next iterations of the design, and then engineered and retested with user-participants, creating a virtuous 'feedback loop' between designers, engineers, and participants.[10]

This interaction between designers, engineers, and participants who are all testers and users of the technology has become our model for collaborative design. I will illustrate the stages and activities of our participant design process. Typically, the lead researcher will first set up an intense design charette, involving other researchers, an engineer, and a designer all working alongside research assistants and student interns. Over two to three days, we aim to discover more about a chosen park location by immersing ourselves in the field. We hope to discover its salient and

6   **F. Lantz,** 'Big Games and the porous border between the real and the mediated,' *Vodafone Receiver*, www.receiver.vodafone.com/16/articles/indexinner07.html.

7   **Multi-modal user interfaces** are a research area in human-computer interaction, providing a user with multiple means of interaction with a system beyond the traditional keyboard and mouse

8   **Drew Hemment,** 'The Mobile Effect,' *Convergence: The Journal of Research into New Media Technologies*, 11 no. 2 (Summer 2005): 32–40.

9   **Also proposed by M. Muller and S. Kuhn,** eds. special issue on participatory design, *Communications of the Association for Computing Machinery*, 36 no. 4 (1993).

10 **Dan Saffer,** *Designing for Interaction: Creating Smart Applications and Clever Devices* (Berkeley: New Riders, 2007).

hidden characteristics, such as natural and cultural features, viewpoints, and typical flora and fauna, as well as more about the human inhabitants, their behaviour, and the nature and locations of their activities in the park. Playing in the space and discovering its potential is an integral part of the research process. Observations and data are collected, and field documentation is created. This documentation often becomes the content for later demonstrations and even final interface designs. An impromptu survey might also be designed and executed. GPS coordinates are set and tested, and then a quick and dirty prototype is improvised, with engineers working directly with designers in the field. Designers and researchers then become testers, often using role play (with sometimes hilarious results), and this is also documented, usually with video and a participant-observer taking notes. Every member of the research team works to their strengths, but they also trade off roles and take on new responsibilities, learning in the field.

When exhaustion or intemperate weather halts play, the researchers decamp back to the lab to compile and compare data, gather and categorize documentation, put it into standardized formats, and integrate findings for the next design iteration. When the new version is ready, and after some preliminary 'bug' testing and 'debugging' in the lab, the same group will return to the field to re-enact the experience, again using role-playing and participant observation but this time equipped with GPS-located devices and mobile phones running the latest iteration of the application. Inevitably, impromptu participation occurs with 'randoms' in the field, who are interested enough to approach us to find out what we are doing, and usually some of them want to play. These unplanned interactions are often extremely revealing and may also be documented with the written permission of the participants. With the clarity of fresh eyes, questions of intention and meaning often arise that we have been staring at too long to see clearly. Constructive misuses of the technology often occur, suggesting that we may have overlooked the obvious in our interface designs. Concepts and ideas are identified and named by our user-participants (with names that often stick), and new possibilities are suggested that wouldn't have occurred to us if we restricted our playtime to the lab and our testing to experienced users. All of the knowledge gathered through the collaboration is rolled back in to further iterations of the design, which are again tested on location, in a cycle of design, test, document, redesign, retest, document some more, refine, retest, etc., until the technology, the application, and the participant experience seem to 'flow' naturally in the park space.

Research and design for mobile experiences must confront and assess issues of usability directly, in this is also best achieved in the intended environment. MDCN researchers Kim Sawchuk and Barbara Crow have developed a research protocol for mobile technologies that draws on standard research methods such as surveys, focus groups, interviews, and participant observation. Their method, however, is distinctively adapted

8 *Park Walk* in High Park
during MDCN Toronto
charette, Summer 2006

9 *Park Walk* in **High Park** during MDCN Toronto charette, Summer 2006

11 **Barbara Crow and Kim Sawchuk** 'Shaking Hands with the User: Principles, Protocols, and Practices for User-integrated Testing in Mobile Design,' in *Mobile Nation: Creating Methodologies for Mobile Platforms* (Toronto: Riverside Architectural Press, 2007).

12 **M. Aurand** 'What is a Charette?' Carnegie Mellon Libraries, www.library.cmu.edu/ Research/ArchArch/Charette/ what.html

13 **Brenda Laurel**, *Design Research: Methods and Perspectives* (Cambridge: MIT Press, 2004).

10–11 **MDCN field research** at Banff National Park

to the mobile realm in that it relies heavily on participatory design principles and 'user-integration' as opposed to the more widely employed user-testing methodologies. Sawchuk and Crow rely on the 'iterative integration of the wisdom of users and designers into the creativity of the research process.'[11] This shift, using classic research techniques integrated with participant observation alongside the participatory and iterative design methods described, moves our mobile research towards a model of designer-participant creation.

Mobile Design Methods and Some Issues of 'Flow'

Researchers, designers, and technologists use a range of approaches for creating new applications in our emerging field of mobile design. Actionable results must be framed by research questions which can be formulated and investigated with participatory design methods and in collaborative formats such as the design charette, in brainstorming and bodystorming sessions, in improvisational techniques, and in user-integrated testing methods.[12] Recent trends in art and design research also encourage workshop activities and the making of artifacts such as collages, paper prototypes, mind maps, and models.[13] Structuring and presentation of the resulting data is a key part of our work as researchers. As designers, we seek to incorporate that data into experiences that will bring delight to our users.

There are particular issues of 'flow' in mobile design in the relationships between the structure of the experience, the functionality of the devices, and their interfaces that, when properly addressed, enable a logical and delightful 'conversation' to occur between the participant, the place, and the device. The spaces and places where users employ mobile technologies are a key factor in their use, and researchers increasingly view the interstices between participants and their locations as an actual interface for careful consideration in the design.

How can technology be employed in such a way that its use augments, and doesn't impede or interfere with, the embodied experience of the park? In urban parks, this involves navigation through areas that may share constituency with dogs and their walkers, tai chi practitioners, skateboarders, children playing, businesspeople lunching, lovers consorting, loafers and itinerants hanging out or meeting on park benches, etc. In our field studies, with the technology conspicuously in hand (the GPS device has to be in a position so that it can always be detected by satellites), we sometimes feel at odds with other inhabitants and users of the park. How can we use technology in easier and less conspicuous ways in the park? Comfort is another issue, whether it is the difficulty of seeing and reading screen displays (we quickly realized that nothing reads on a dark background in bright light, and that text was not really a good option) or distinguishing particular levels and sorts of audio in the outdoors. And while Bluetooth works through a pocket lining, how do you transport that GPS device, making certain it is constantly in the open? What happens when you give it to someone else, and break the invisible,

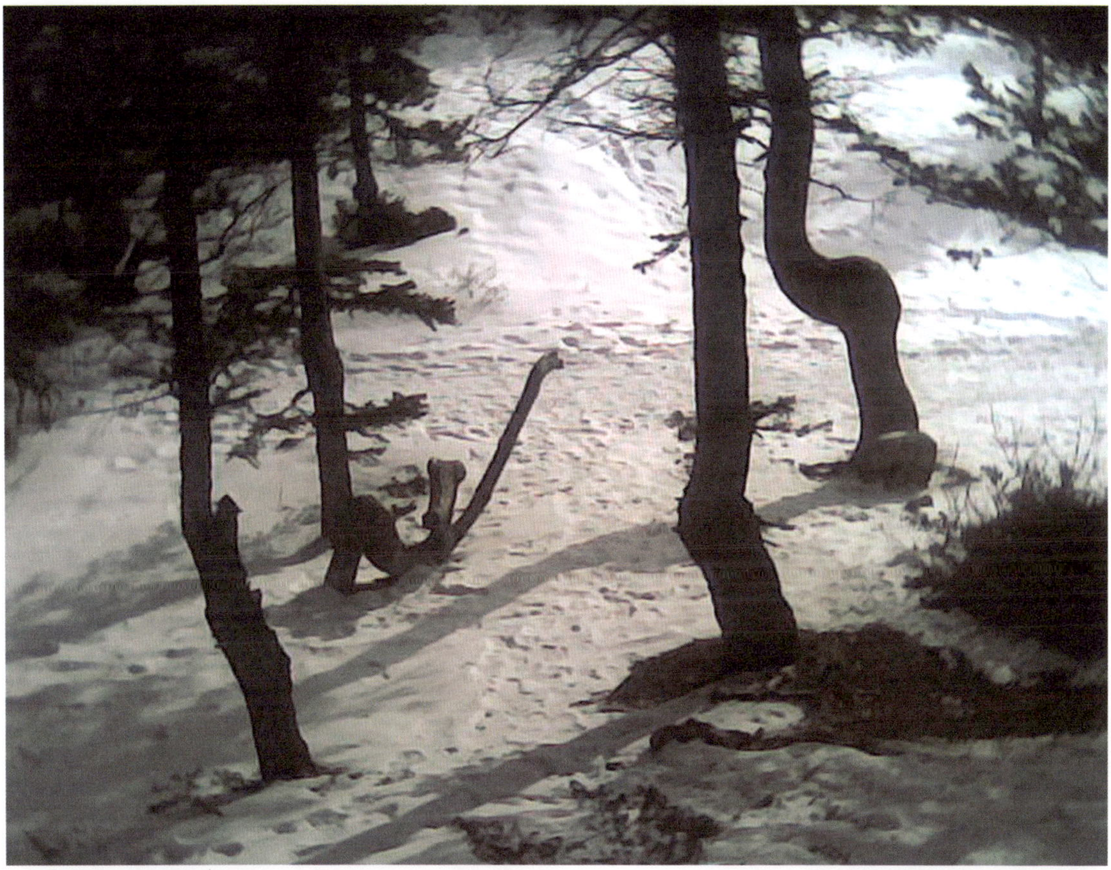

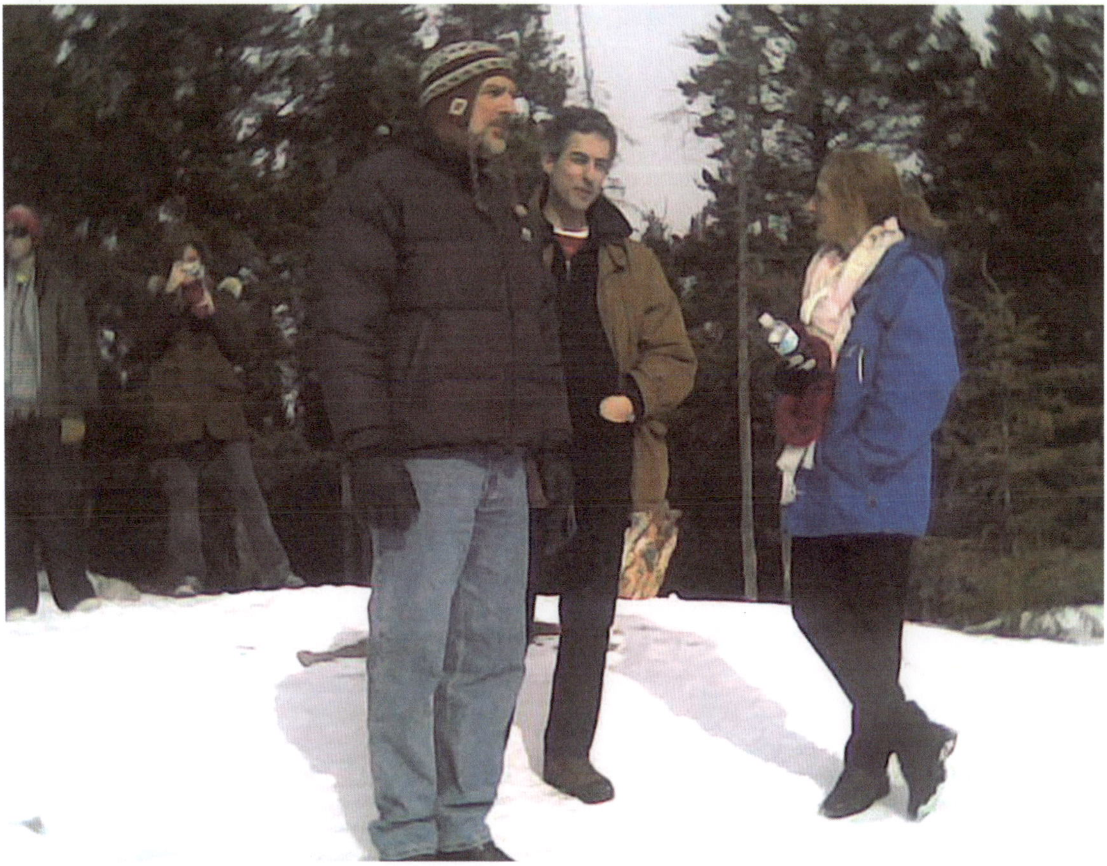

magical ten-metre Bluetooth signal tether? Ingenious accessories and garments with mobile and GPS carriers, Velcro body attachments, gloves, and even GPS walking sticks, hats, and helmets were all imagined, with some were prototyped and tested within the MDCN.

Safety is an important issue for park visitors in both urban and wilderness locations; this is a prime design consideration. How do we design a technologically augmented experience in ways that will not compromise the safety of participants or other users of the park? Our designs need to be engaging, but not so complex or immersive that inattention to surroundings results. Apart from getting lost in virtual space and losing contact with the physical locale, dangerous proximity to cliffs, raging rivers, and local wildlife are all issues we encountered. Whether to use earbuds with the experience was a consistent subject for debate that was finally settled when a headphone-wearing cyclist on the Hoodoos Trail in Banff was injured in an encounter with a black bear, probably because he couldn't hear what was happening on the trail ahead. Luckily, this didn't occur in one of our field trials!

Weather is also a major factor, and our field research in Banff and Mount Royal Park in midwinter bears witness to the fact that all aspects of the experience of the urban and wilderness parks of Canada are extraordinarily influenced by seasonal weather variations. Fingers in heavy mittens don't navigate handsets well, and mobile device batteries die quickly in subzero temperatures. A deep fall of snow will completely change both the look and feel of the environment (essentially the 'set' for the experience) and affect the visitor's interest and ability to navigate and perform outdoor activities. Designing season-specific applications is one way to meet this challenge.

'Indispensable' Devices

The concept of creating a guided, immersive experience in the outdoors is not novel, and neither is the idea of providing a technologically or mechanically enhanced guide to the traversed landscape. This was wonderfully illustrated in the late nineteenth century in Edward Bellamy's highly imaginative and prophetic writing about 'indispensable' devices.[14] In his 1898 short story *With the Eyes Shut* the protagonist on a train journey is treated to the experience of what could only be described as a 'talking book.' Upon expressing great satisfaction with the experience, the 'train boy' tells him of an even more glorious invention:

> *In reply he volunteered the information that next month the cars for day trips on that line would be further fitted up with phonographic guide-books of the country the train passed through, so connected by clock-work with the running gear of the cars that the guide-book would call attention to every object in the landscape, and furnish the pertinent information—statistical, topographical, biographical, historical, romantic, or legendary, as it might be—just at the time the train had reached the most favorable point of view.*

12 *Park Walk* in Grange Park, Toronto

13 *Park Walk* **researchers** in High Park during MDCN Toronto charette, Summer 2006

14 **E. Bellamy**, 'With the Eyes Shut,' in *The Blindman's World, and Other Stories.* (Elibron Classics: 2001, originally published 1889).

The *Park Walk* project addresses this long-familiar desire to find out more about our surroundings in environments unencumbered by maps, signs, or plaques. Mobile technologies can provide visitors with located information, stories, and entertainment in situ, as if pulled from the air. They also encourage visitors to create, capture, and collect their own stories.

Bluetooth and GPS technologies have unique characteristics that allow participants to be 'sensed' when strolling through a location, triggering the push of information and invitations to interaction within the environment via the mobile phone. In one of the *Park Walk* prototypes, participants are oriented and encouraged to freely explore Grange Park, adjacent to the Art Gallery of Ontario and the Ontario College of Art & Design, where they are given an immersive audiovisual historical/cultural tour of the park and its surroundings, buildings, and features. Mobile devices are used in a unique way in the *Park Walk* project, in part as technologies of the gallery, usurping the traditional mechanics of display, reconstruction, and simulacra.[15] Taking a cultural historical conversation out of the museum setting and placing it in the park breaks the bonds of institutional space and places the display in an interactive, social setting. Hence, mobile technologies can enable an accessible experience for a wider and more diverse audience. The *Park Walk* project situates and ties information directly to place and, by inviting contribution, creates a narrative space that is open to collaboration, enabling deeper interaction between the visitor, the community, and the park.

[15] **Gillian Rose**, *Visual Methodologies: An Introduction to the Interpretation of Visual Materials* (London: Sage Publications, 2001).

Acknowledgements

I am indebted to all of my MDCN colleagues for their research and support, but especially to my collaborator Bruce Hinds; engineers Tom Donaldson, David Gauthier, and Jagmit Singh; designer Nevena Niagolova; research assistant Ken Leung; and OCAD student interns Bryn Reed-Ludlow, Janet Bewell, Amanda Cooley, and Garry Ing.

Adventures in Mobile Culture Media

David Vogt
University of British Columbia

1 **Mobile MUSE** focuses on mobile-enabled social interactions

The Mobile Media-rich Urban Shared Experience (MUSE) Network was conceived in 2003 to cultivate a British Columbia mobile media industry in preparation for the 2010 Winter Olympics. Pioneering MUSE mobile culture media applications include *Mobile Vancouver International Film Festival (Mobile VIFF)*, *metroCode*, *English2Go*, and *Pocketcine*, collectively a central motivation behind Vancouver's February 2007 decision to deploy municipal Wi-Fi by 2010. The Mobile MUSE Network carries forward invaluable know-how, momentum, and key assets including (a) a proven innovation model based on public prototyping of mobile media technologies within projects bringing together multi-sector champions (academic, corporate, community, cultural, and new media leaders); and (b) a versatile mobile services platform for bidirectional SMS, MMS, and VoiceXML applications, along with a broad suite of tools and supporting technologies.

The original MUSE acronym (Media-rich Urban Shared Experience) remains relevant to a dramatically new focus on community-generated media (CGM) to enliven open public spaces ('live spaces').

Interactive public displays will become ubiquitous over the next few decades. Broadband wireless networks will link giant screens in public places with large screens in nearby pubs, malls, and homes, as well as tiny screens in everyone's hands. There are extraordinary cultural potentials for the collective media experiences that can be staged through these screen networks. Such experiences will be interactive, participative, distributed, and embedded in or tempered by local contexts—creating unprecedented 'community centres' and 'channels' for cultural expression, community development, and economic diversification. CGM will be hyperlocal and everywhere. The 2010 timeline allows Canada to assert a creative leadership position at the convergence point of these emerging technologies.

Live sites and CGM provoke vital questions to inspire new media research:

- What program and event technologies effectively blend content and interactions from many sources?
- What flow dynamics govern experiences staged across networks of giant, large, and tiny screens?
- How can communities own this new channel and use it to foster identity and prosperity?

Leading up to 2010, Canada will install perhaps the largest network of live sites in the world, yet we don't have the native experience, know-how, or insight into how these installations can become more than large television sets. CGM is the user generated content (UGC) phenomenon raised to community scales—an essential solution for creating indigenous content and Canadian identity, and crossing digital divides.

facing page

2 **The wellspring** for mobile experience is cultural expression

3 **A 'mobile nation'** can inspire a 'mobile planet'

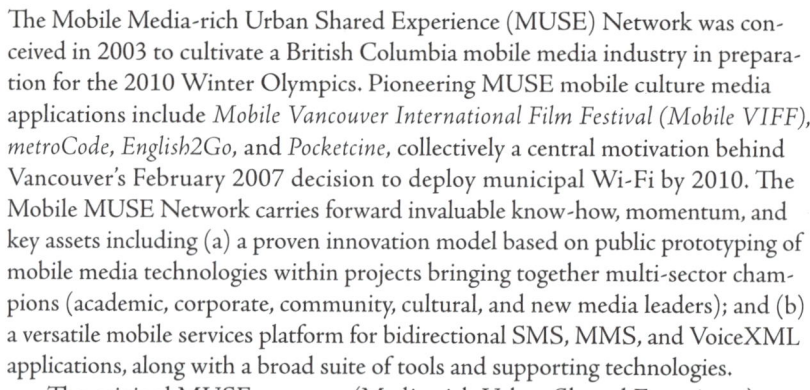

your idea CONNECTS
people with URBAN CULTURE

join the GLOBAL CONVERSATION

mobile
muse.ca

Day of the Figurines
a pervasive game for SMS

Matt Adams
Blast Theory

Day of the Figurines is a massive multiplayer game for mobile SMS (short message service). Players enter the game by visiting a large metal model of an English town and placing their plastic figurine into the model.

The game then unfolds over a total of 24 days, each day representing an hour in the life of the grimy little town as it shifts from the mundane to the cataclysmic: the local vicar opens a summer fête, Scandinavian metal musicians play a gig at the Locarno bar, and an Arabic army appears on the high street. Up to a thousand players respond to these events and to each other as they roam from the Gasometer to the Product Barn, from the canal to the Rat Research Institute.

Matt Adams joined us at Mobile Nation via video conference, and reflected from his home in Brighton, UK, on the pervasive elements of the game, discussed game design issues for mobile devices such as pacing and message aggregation, and illustrated ways in which cultural differences in SMS use have been manifested in games in Berlin and Singapore.

Blast Theory created *Day of the Figurines* with collaborators in the Integrated Project for Pervasive Gaming, a three-year research project funded by the European Union.

1 **Uncle Roy All Around You**
Interactive game played online and on the streets of London using handheld computers, 2003

facing page
2 **Day of the Figurines**

3 **Can you see me Now?**

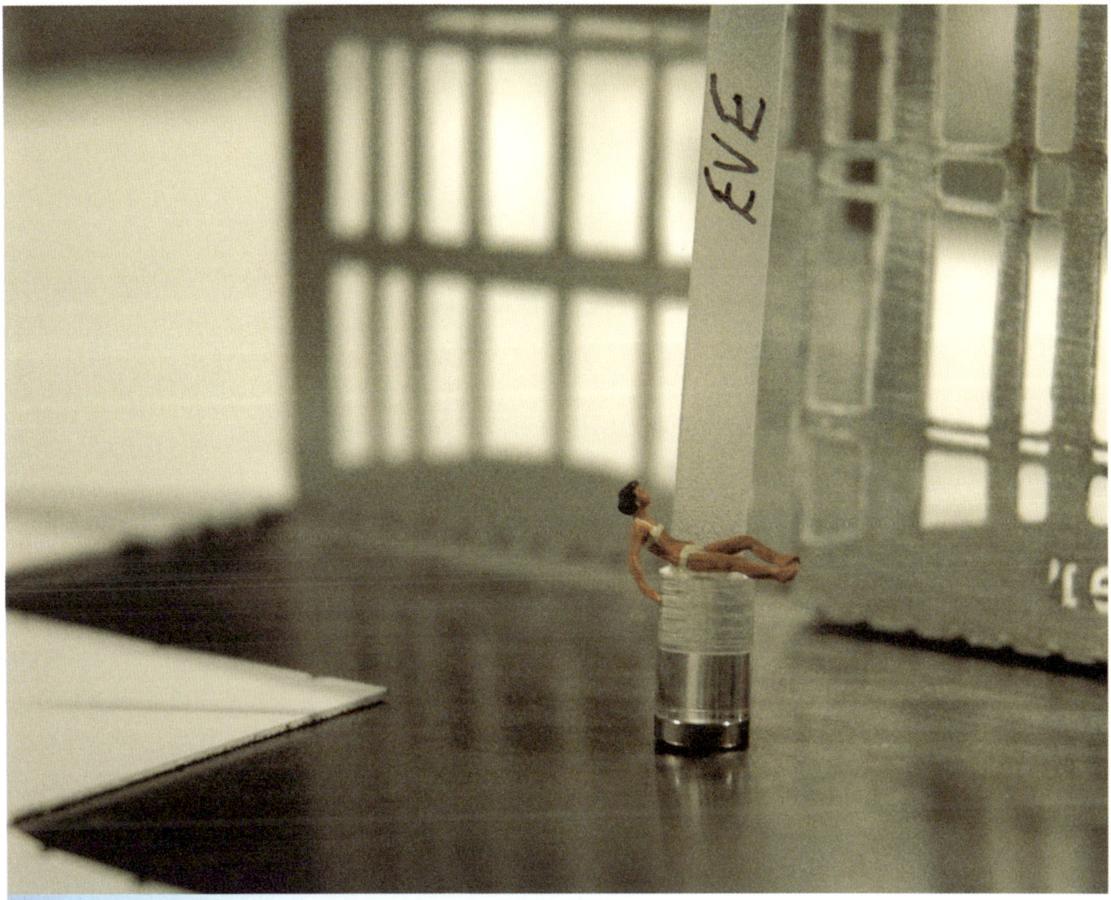
EVE

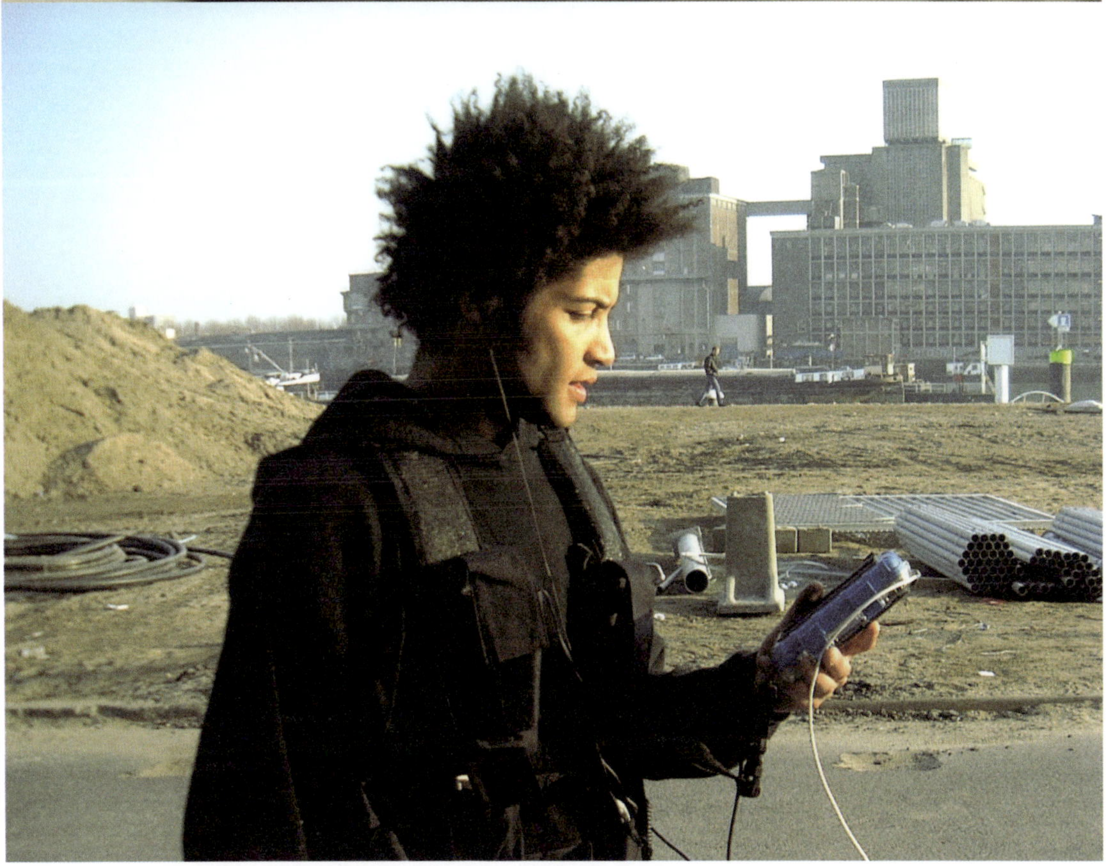

Electronic Textiles and Reactive Garments

Joanna Berzowska
Concordia University

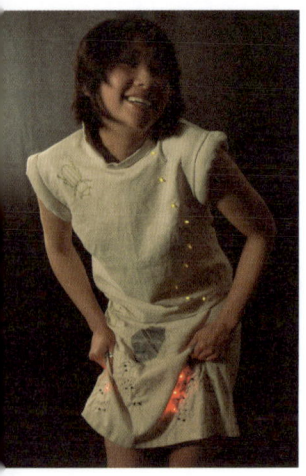

I **Intimate Memory**
This garment consists of a shirt and a skirt which employ two different input and output methodologies, including LED light intensity, to record acts of physical intimacy and indicate the time elapsed since those 'intimacy events' have occurred

facing page
2 **Detail** of Intimate Memory

3 **Constellation Dresses**
The Constellation dresses are covered with roughly twelve magnetic snaps irregularly arranged over the torso and thighs and connected in pairs; LEDs are integrated into the dresses in a design that resembles a constellation, with a cluster of stars connected to each other through short and straight lines

The developed world is becoming increasingly dense with electronic devices. Our power consumption needs are constantly increasing, particularly in the case of mobile and wearable electronic devices, and current trends indicate this will continue to be an issue in the future. Since the development of alternative energy sources has not yet yielded economically viable solutions and has not kept up with the needs of our emergent and expanding markets, we are heading towards an environmental disaster.[1] At the same time, one of the driving forces for fashion, spanning history and different cultures, has been to seek a continually evolving concept of beauty through the transformation of the body's natural form. This has been exemplified by practices ranging from subtle adjustments of a body's proportions through the use of conical brassieres, bustles, crinoline hoops, and exaggerated shoulder pads, to more extreme practices such as deliberate (and sometimes permanent) physical deformation of the body.[2]

Captain Electric and Battery Boy is the code name for a new XS Labs collection of garments that directly address issues of power consumption and sustainability by creating body-worn, textile-based 'living organisms.' We are creating kinetic electronic garments that harness power from the body and use that energy to transform themselves in response to various internal and external stimuli.

We propose to develop three wearable artifacts, called *Sticky*, *Itchy*, and *Stiff*, that constrain body motions in a variety of ways. Responding to the need to address long-term sustainability in new technology development, *Sticky*, *Itchy*, and *Stiff* will both passively harness energy directly from the body and actively allow for power generation by the user. Depending on levels of discomfort and extenuation, as well as the desire to supersede the limitations of the human body, the garments will produce varying amounts of energy to fuel themselves. This is called 'parasitic power.' The garments will then use that power to move and change shape on the body, using embedded Nitinol (shape memory alloy) fibres.

XS Labs is a design research studio based in Montreal where we develop electronic textiles, wearable computing, and reactive garments. We are concerned with the exploration of simple interactions that emphasize expressive

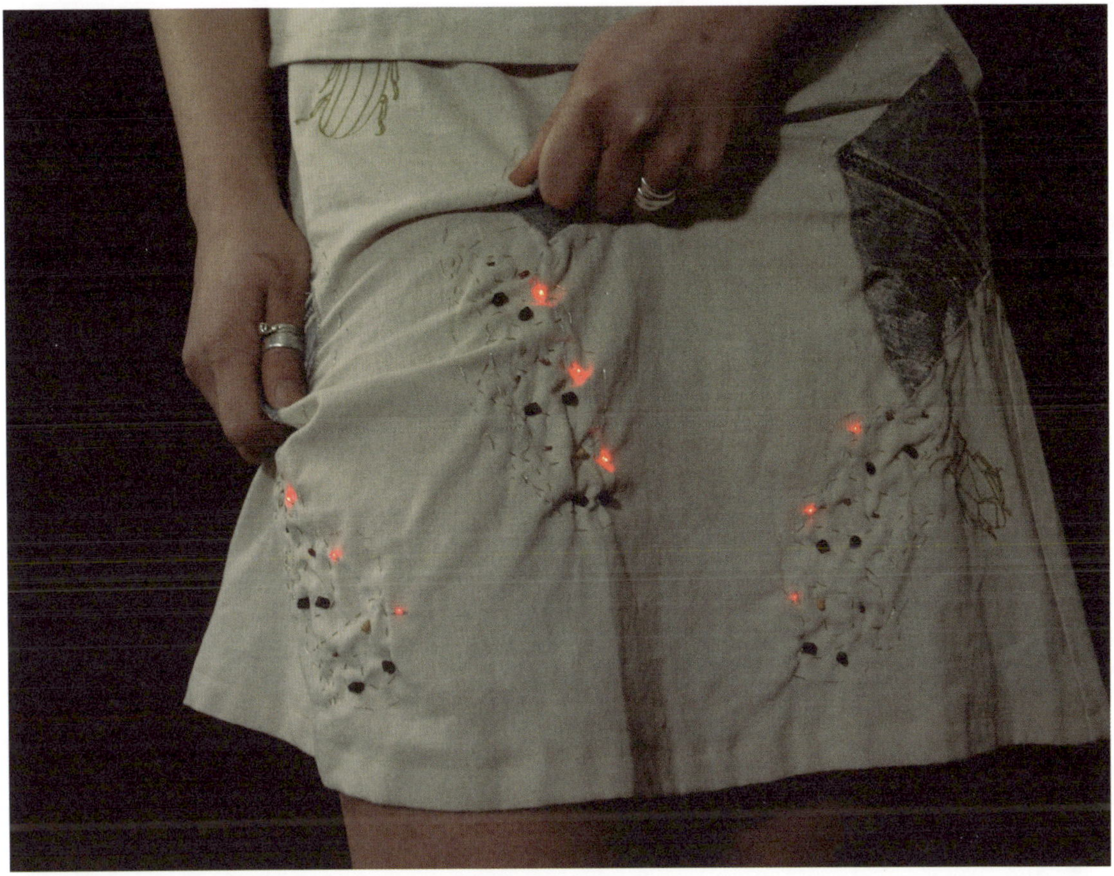

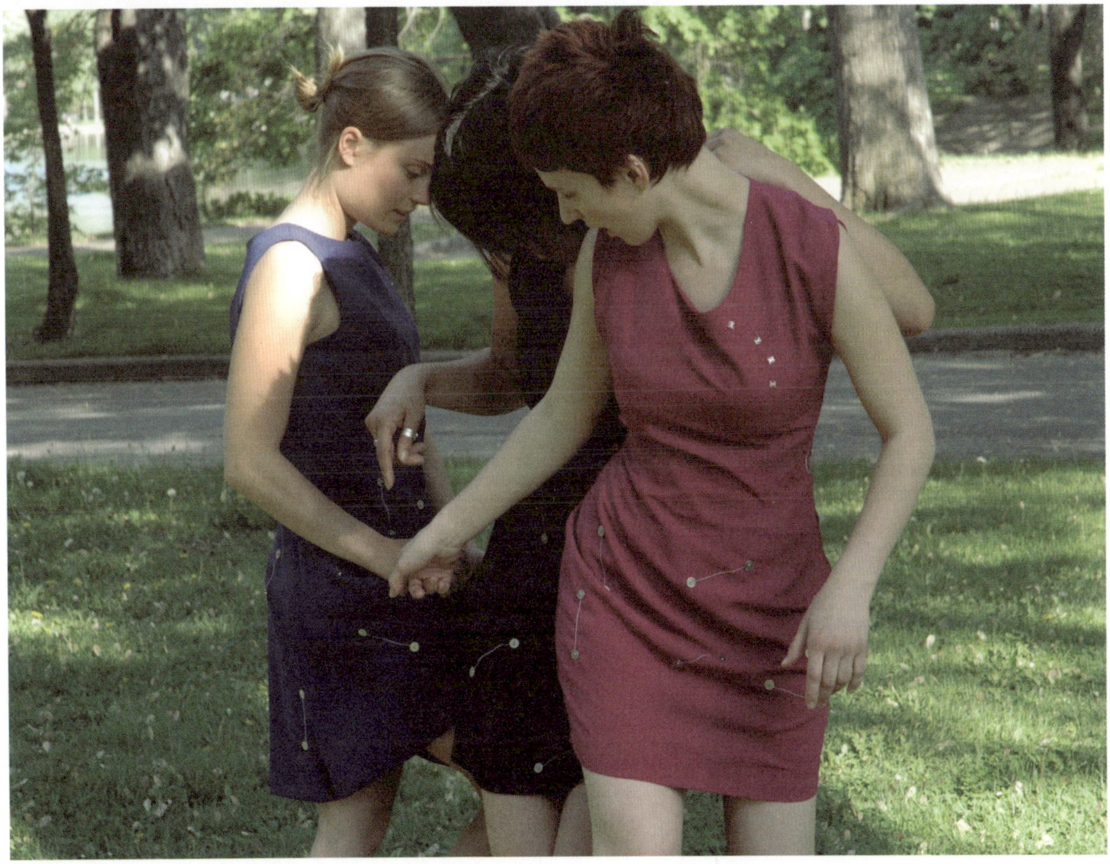

qualities of electronic circuits and of the body. Of particular interest to XS Labs are the many relationships between our bodies and the architectural spaces that they inhabit. Our clothing is one of the first such structures, often talked about as a 'second skin,' which enables an important level of interface between the human flesh and the outside world, physically and metaphorically. This is why we are concerned with active materials that can be easily integrated into textile substrates and that can be controlled through soft electronics.

An important research direction at XS Labs attempts to solve technical problems and look at new construction methods for the development of textile substrates that function as soft electronic circuit boards. We construct simple electronic components with techniques such as weaving, sewing, embroidering, and tying knots. Soft electronics are important, since wearable technologies are intrinsically close to the body and need to be comfortable and even pleasurable to wear.

Notes

[1] T. Starner and J.A. Paradiso, 'Human-Generated Power for Mobile Electronics,' in ed. C. Piguet, *Low-Power Electronics Design* (Boca Raton: CRC Press, 2004), 1–35.

[2] H. Koda, *Extreme Beauty: The Body Transformed* (New York: Metropolitan Museum of Art, 2001).

4–5 **Kukkia Kinetic Electronic Garment**
The Kukkia flowers frame the face and slowly open and close over time, like a caress. The dress does not respond to proximity, mood, or the stock market. Rather, it is an expressive and behavioural kinetic sculpture that develops a visceral relationship with the wearer. The Kukkia flowers are constructed out of felt and silk petals that provide relative rigidity and integrate stitched Nitinol (shape memory alloy) wire, which enables the slow, organic movement.

facing page
6–7 **Vikas**
Vilkas is a dress with a kinetic hemline on its right side that rises over a thirty-second interval to reveal the knee and lower thigh, creating a kinetic dress whose behaviour can be playful, even desirable, but also contextually embarrassing

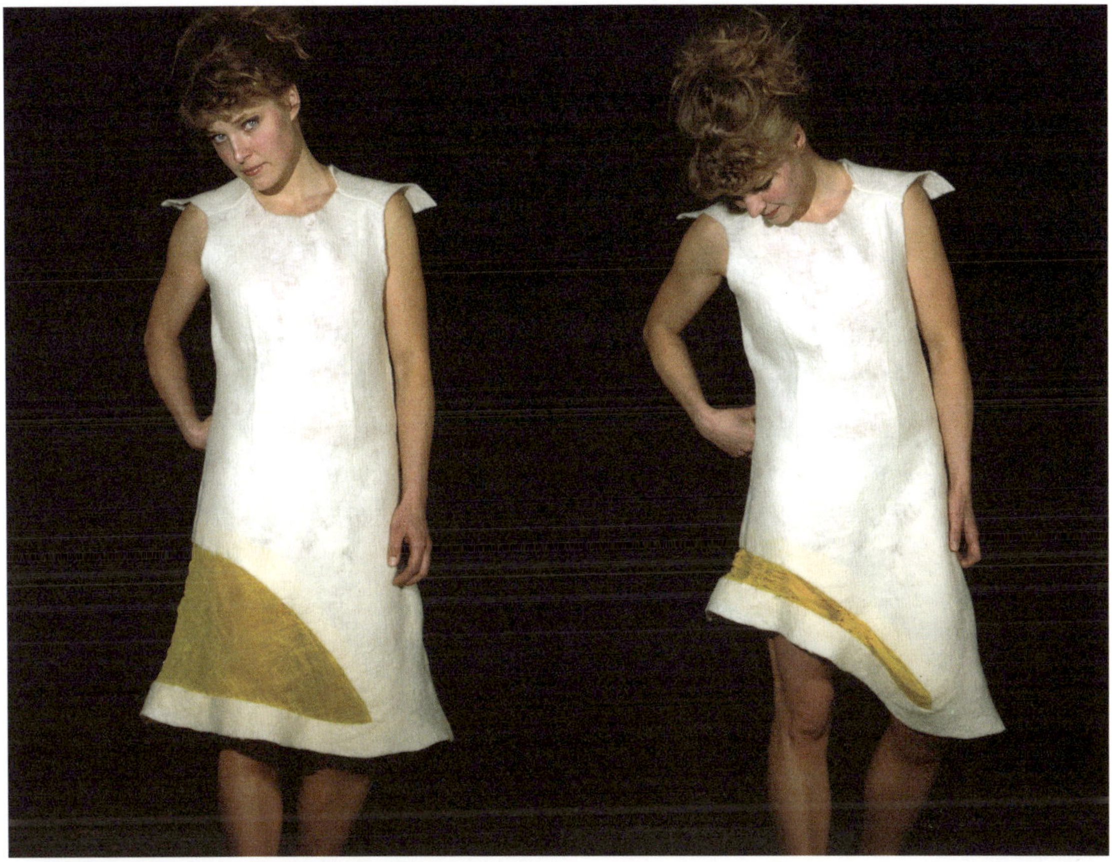

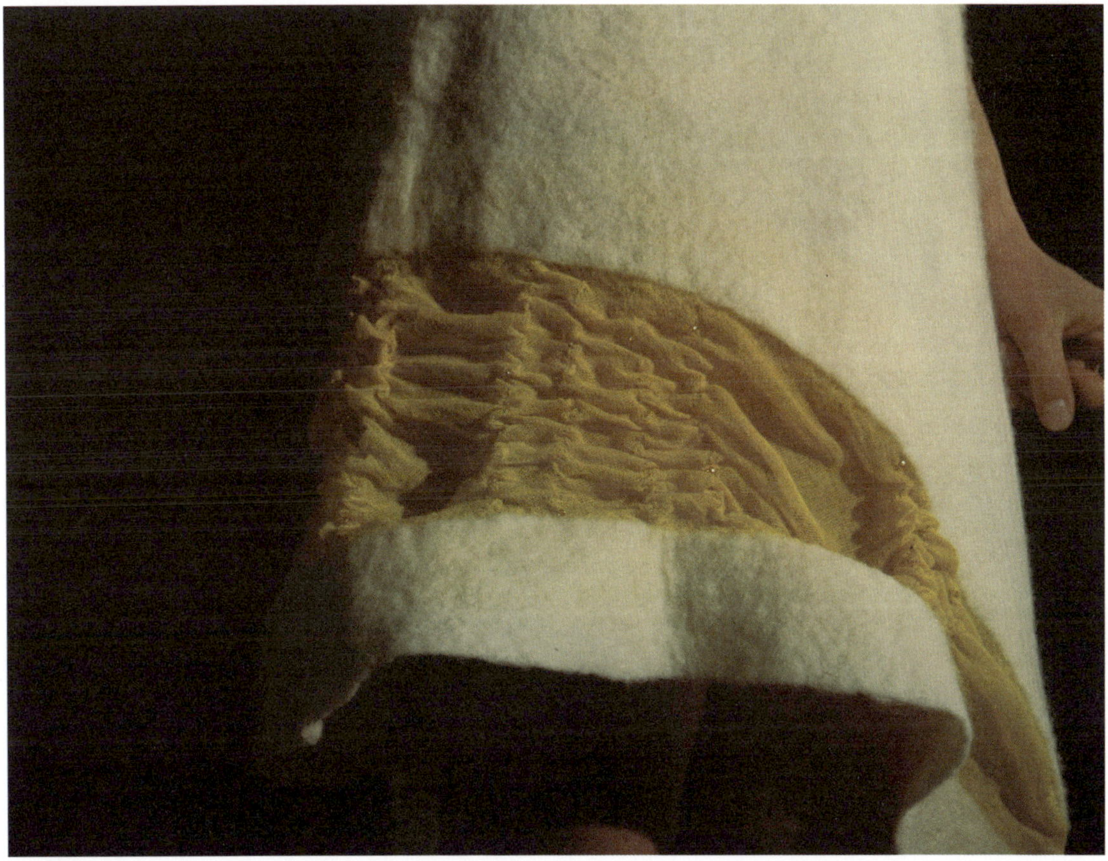

(soft^n)
survival strategies for interaction

Thecla Schiphorst
Simon Fraser University

This paper describes and contextualizes a new interactive artwork created in collaboration with V2_Lab in Rotterdam that explores qualitative strategies for interaction, communication, and a notion of 'group intimacy' through a tangible networked ecology of soft objects that exhibit emergent properties of behaviour and that are 'interfered with' through participants and users of the system.

What?

(soft^n) is a working title for an interactive public art installation based on exploring emerging network behaviour through interaction between a group of soft networked objects. It takes place in a social urban setting, such as a café or a lounge. One can think of (soft^n) as a counterpoint to, or a critique of, the hard: a survival strategy for interaction that allows misplaced action, mistakes, forgiveness, bad attitudes, weakness, stillness, giving in, and throwing away. (soft^n) evokes critique through the computational act of quality.

Huh?

And quality is a form that ruptures binary conditions; it is fuzzy, it adds (+) more-than-1 (>1) logic to a space, a space of many logics, or at least a form of perceptual quality to the system: computational quality. The play between soft and hard elicits and describes a set of qualities which speaks to methodologies, design strategies, form factors, and interface paradigms within technological development. And as our form factors dissolve or give up (their hardness), (soft^n) attempts to make room for a space that contains a 'willingness' to explore audience interference, mistaken identity, and a socially liminal intimacy.

Exploring qualitative aspects of the connection of a body/object to itself, to another, and the concept of group-body/object, a group's ability to sense 'data exchange' and to amplify, intervene, or augment that exchange through choice, attention, and physical manipulation.

Where?

Situated in a café, club setting, or speakeasy, this experience uses the fabric of urban spaces, urban environments, and social interactions as a site for remixing itself. (soft^n) can be thought of as a kind of urban choreography.

facing page
1–2 **(soft^n)**
Interactive soft object 1

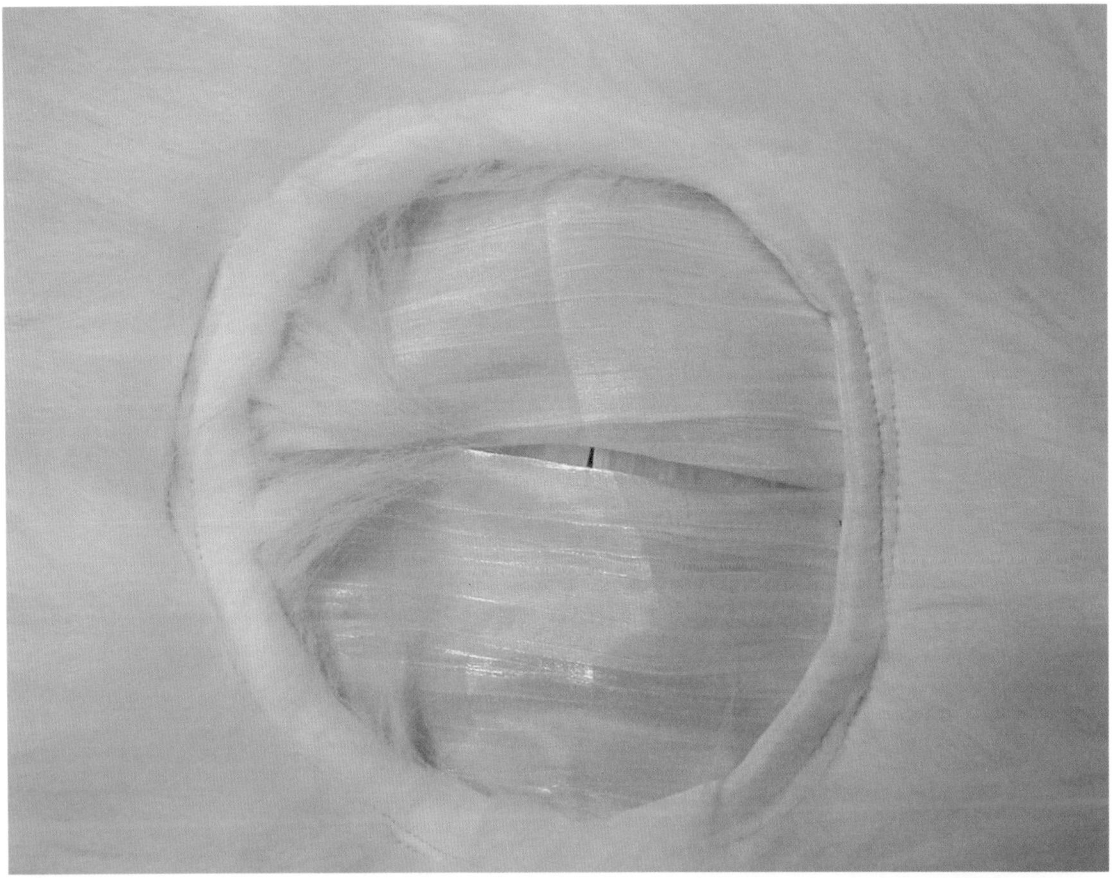

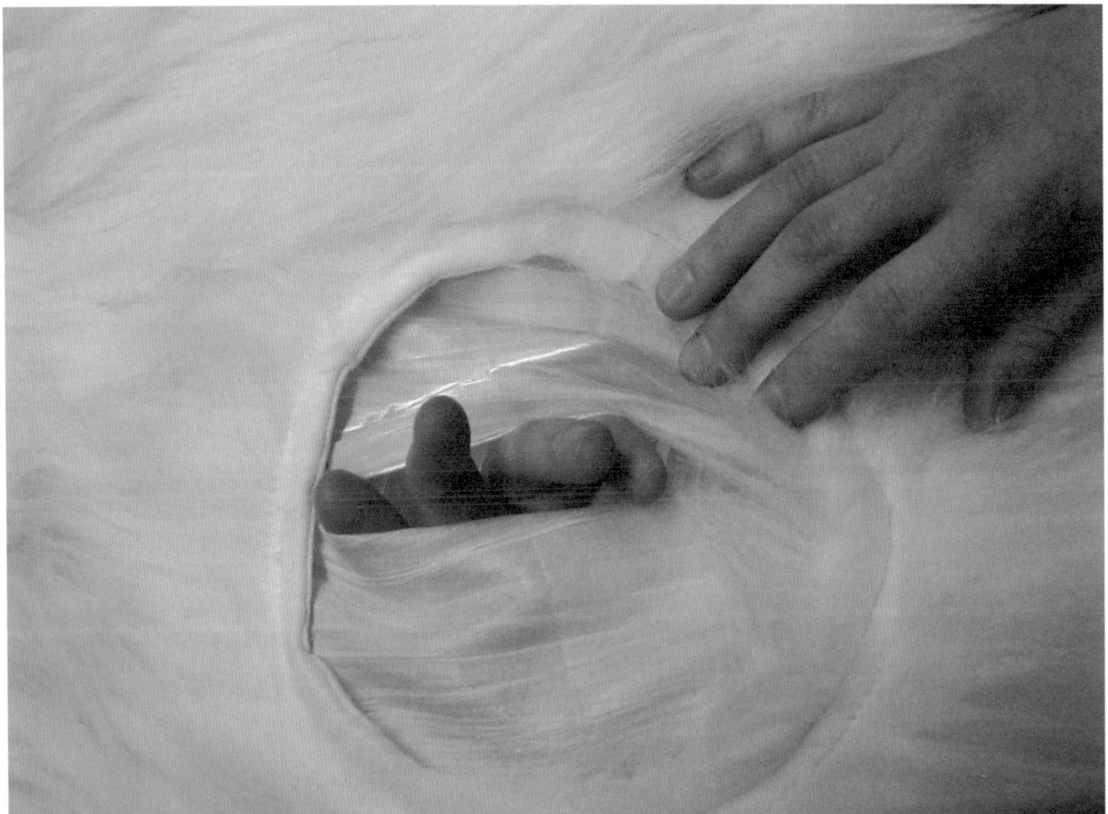

How?

(*soft^n*) utilizes a group of soft objects that are strewn about, and tumbled within, a public urban space, and that are networked to one another, that create a group-body based on body-data of participants: touch, movement, and physical data. The (*soft^n*) objects create an ecosystem where the audience interferes and where new behaviour comes into existence through audience interaction.

The design includes a group of twelve interactive soft objects, each containing a specially designed and custom-engineered multi-touch soft input surface, motion detectors, an ability to output movement (vibration), light, sound, and physical deformation, and an ability to communicate wirelessly to each other within the networked group.

This piece includes the development and testing of an interaction model based on input heuristics of touch and movement: the basis for developing exploratory behaviour within a network ecology that allows and supports notions of audience interference, rules of control, open scenarios, survival strategies for interaction, and the soft objects' evolution of their own 'best practices' through their interaction lifecycle of attention, quality recognition, and use.

And How Many?

The *n* in (*soft^n*) represents multiples: multiple nodes, multiple objects, multiple iterations, multiple participants, multiple experiments. And of course *n* is for network, and the notion of the variable which suggests emergence, emergent behaviours, and multiple frameworks—a critique of Human Computer Interaction (HCI), a domain within computer science that studies user behaviour, response, and interface, that also uses HCI as a method to explore artistic practice, a bridging methodology, a user interference.

facing page
3 **(soft^n)**
Interactive soft object 2

4 **(soft^n)**
Interactive soft object 3

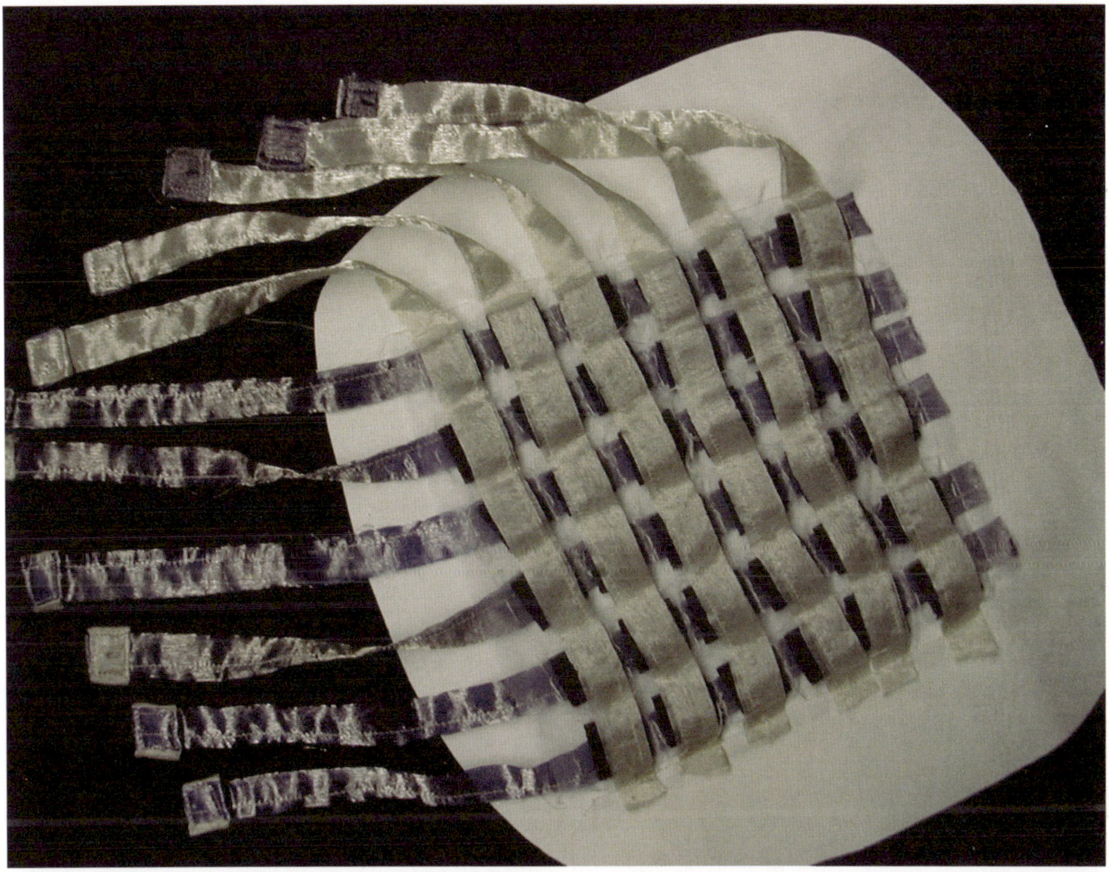

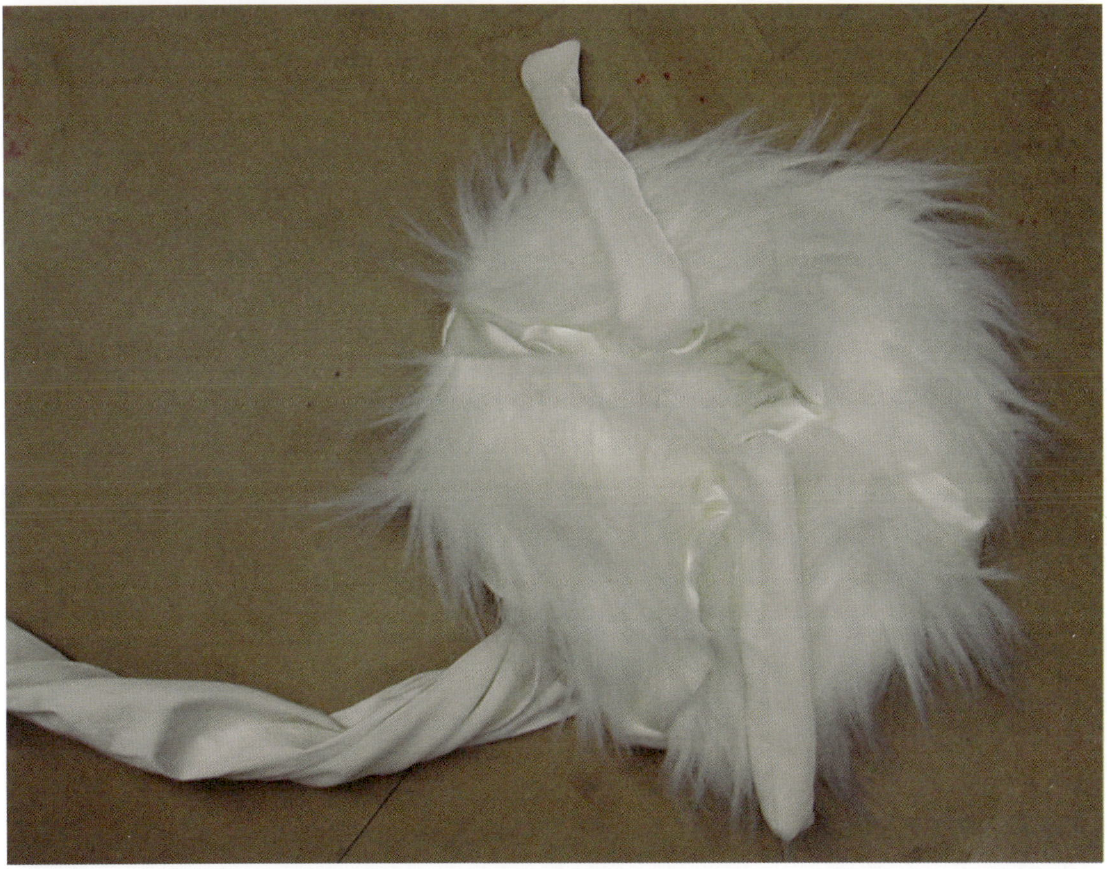

Malleable Matter
adaptable and responsive space

Filiz Klassen
Ryerson University

This paper is an investigation of the recent research and developments on high-performance textiles, smart textiles, and hybrid materials. The term 'high-performance' connotes the designed or enhanced properties that improve the materials' performance in specific conditions but stay fixed or static in response to external stimuli. The term 'smart' or 'intelligent' refers to materials that change their properties in response to varying thermal, luminous, acoustic, or structural stimuli. Although the terms 'fabric' and 'textile' are used in construction and resemble the properties of cloth with natural fibres (such as cotton, wool, and silk), high-performance or smart textiles are engineered with synthetic fibres (such as nylon, polyester, carbon, and glass), special coatings, embedded technology, sensors, and electronics. Many other composite materials that are flexible and layered are termed 'hybrid materials' as they share properties of the originating materials such as textiles and plastics. These materials have the potential for weaving a new direction towards materiality in design and construction.

Adaptability/Responsiveness:
Shift in Perceptual Boundaries

I believe that the material landscape in built environments is currently in a state of transition in which our current design practices will inevitably be transformed in a direction compatible with the theme of responsive environments. Recent high-performance and smart material innovations are demonstrating a new kind of adaptability and transformability of space that is different than those that simply focus on mechanical expansion of spaces and objects. This phenomenon perhaps can be best explained with Philip Ball's concept that smart '[m]aterials can replace machines'. He suggests that '[s]ubstances that change their shape or properties in response to various stimuli—electrical signals, light, sound waves—can be used as switches and valves with no mechanically moving parts.'[1] In the hands of material scientists this may mean the development of certain types of advanced materials that, for example, not only give warnings of structural malfunctioning (by colouration or discolouration of material) but also counterbalance, amorph, or provide temporary strength to deflect external or extreme forces, such as high winds or earthquakes.

[1] P. Ball, *Made to Measure: New Materials for the 21st Century* (Princeton: Princeton University Press, 1999), 103.

facing page
1–2 **Reflective fabric microphotography**

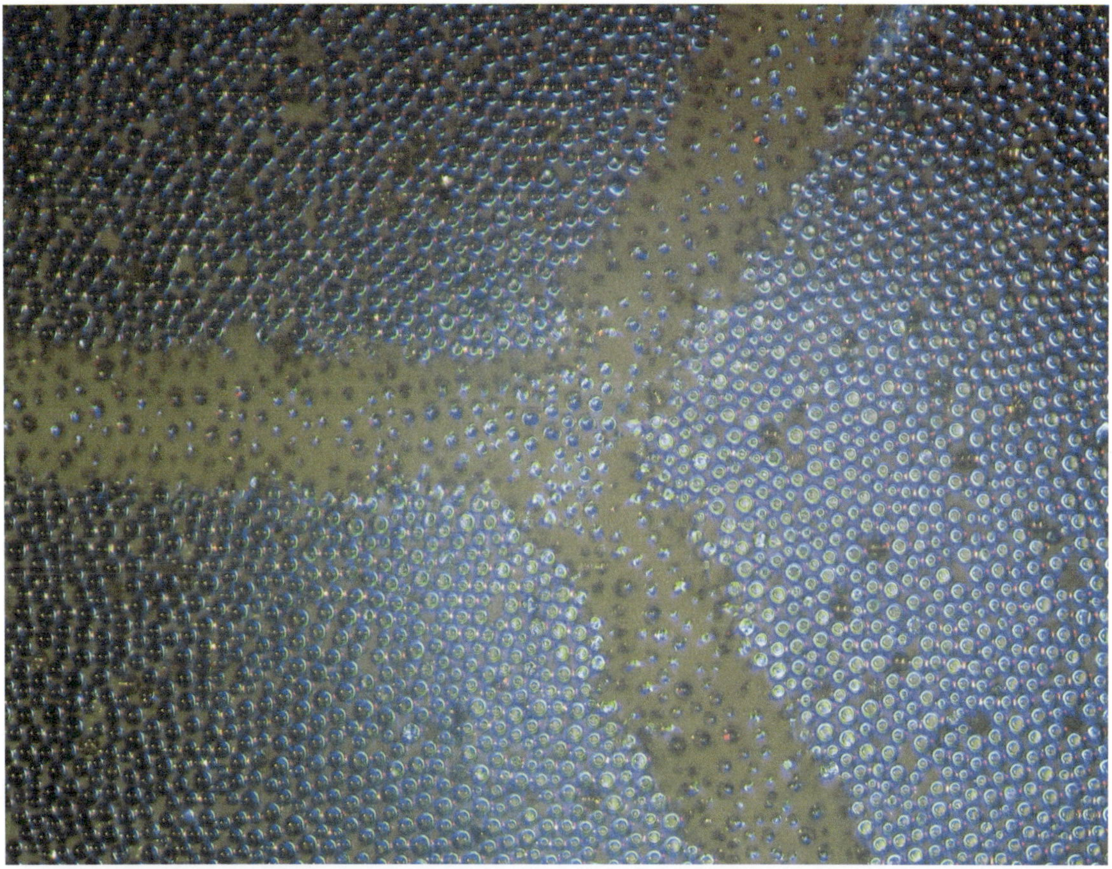
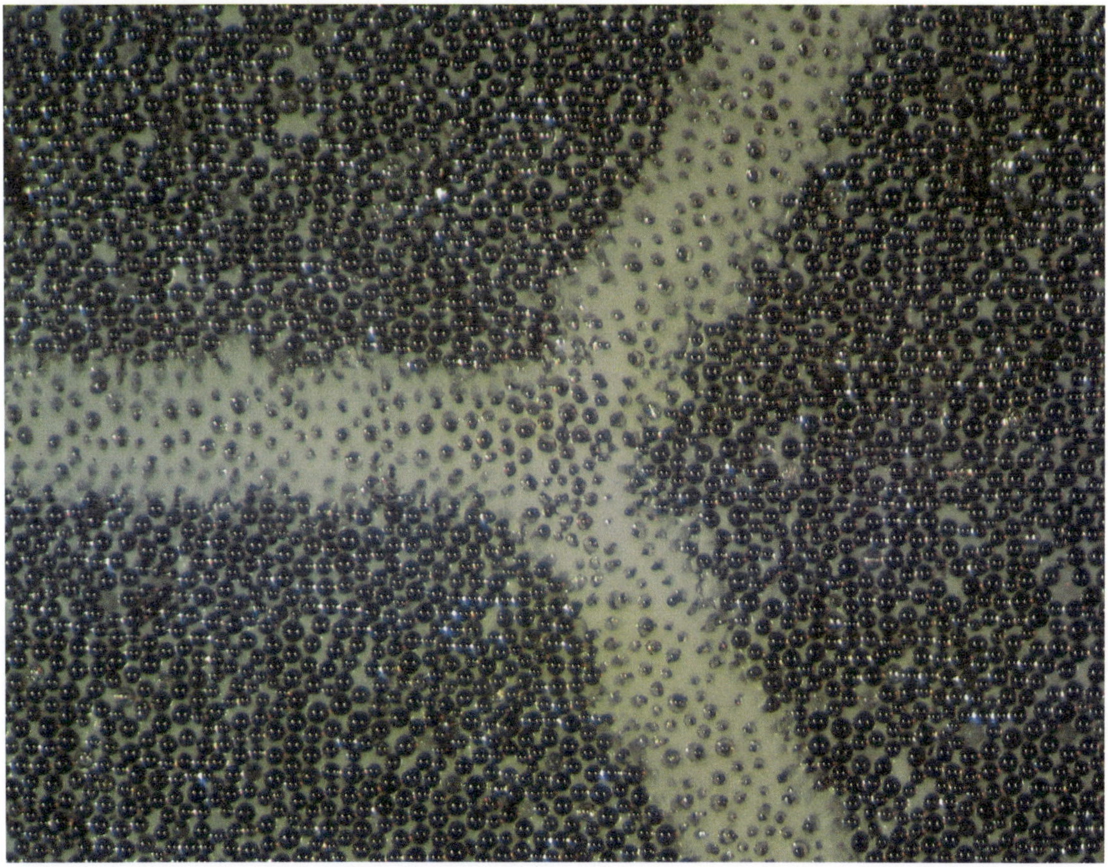

The integration into architecture of materials that change their static state through deformation, re-formation, or even destruction under stress or temperature differences presents a scientific as well as a design challenge.

> *In the past, a change in a material's properties (its elasticity, or its volume) in response to a change in the environment was generally seen as a potential problem, as a thing to be avoided… Even in applications where one might imagine a dumb [static] material would suffice, a degree of smartness may prove tremendously useful… A house built of bricks that change their thermal insulating properties depending on the outside temperature, so as to maximize energy efficiency?*[2]

2 **Ball**, *Made to Measure*, 104.

New Directions in Materials Research and Architecture

Scientific research is continually improving mechanical, thermal, electrical, chemical, and optical properties of materials in architecture. Therefore it seems obsolete to choose materials solely based on their visual characteristics in conception and creation of spaces. What becomes relevant is to find out what a material might do in a space to enhance our all-sensory experience.

In his book *Made to Measure*, Philip Ball predicts that there will be always room for so-called dumb (i.e. static) materials that do not change their properties or display their changing characteristics. Nevertheless, it will increasingly pay to be 'smart' in the manner discussed above, although he maintains that this is still not nearly sufficient. In the future, material scientists hope that materials will be developed that are able to take into account changes, maintaining 'a memory of what has transpired before and that learn from these previous experiences' and becoming 'more active' and 'smarter' or 'more intelligent' as they get older.[3] He further comments that in the 1995 aircraft prototype developed by researchers at Auburn University,

3 **Ball**, *Made to Measure*, 105.

> *all of the ailerons and tail flaps that are used to control the flight of conventional aircraft were replaced by wings and tail fins containing piezoelectric actuators (that convert electrical to mechanical energy) that altered their shape (in response to flying conditions). One advantage of smart wings is that they can be continually adapted to maximize aerodynamic (and thus fuel) efficiency in a way that is just not possible for today's aircraft'*[4]

4 **Ball**, *Made to Measure*, 118.

Michelle Addington and Daniel Schodek, authors of *Smart Materials and New Technologies for the Architecture and Design Professions*, suggest that 'by investigating the transient behavior of the material, we [can] challenge the privileging of the static planar surface' that long dominated the architectural vision. They further propose that '[s]mart materials, with their transient behavior and ability to respond to energy stimuli, may eventually enable the selective creation and design of an individual's sensory experiences.'[5] Although they conclude that architects are not in a position to exploit this alternative paradigm shift in material innovations, by examining the knowledge gained from other industries such as aerospace

facing page
3–4 **Fabric with fibre optic light emitter**

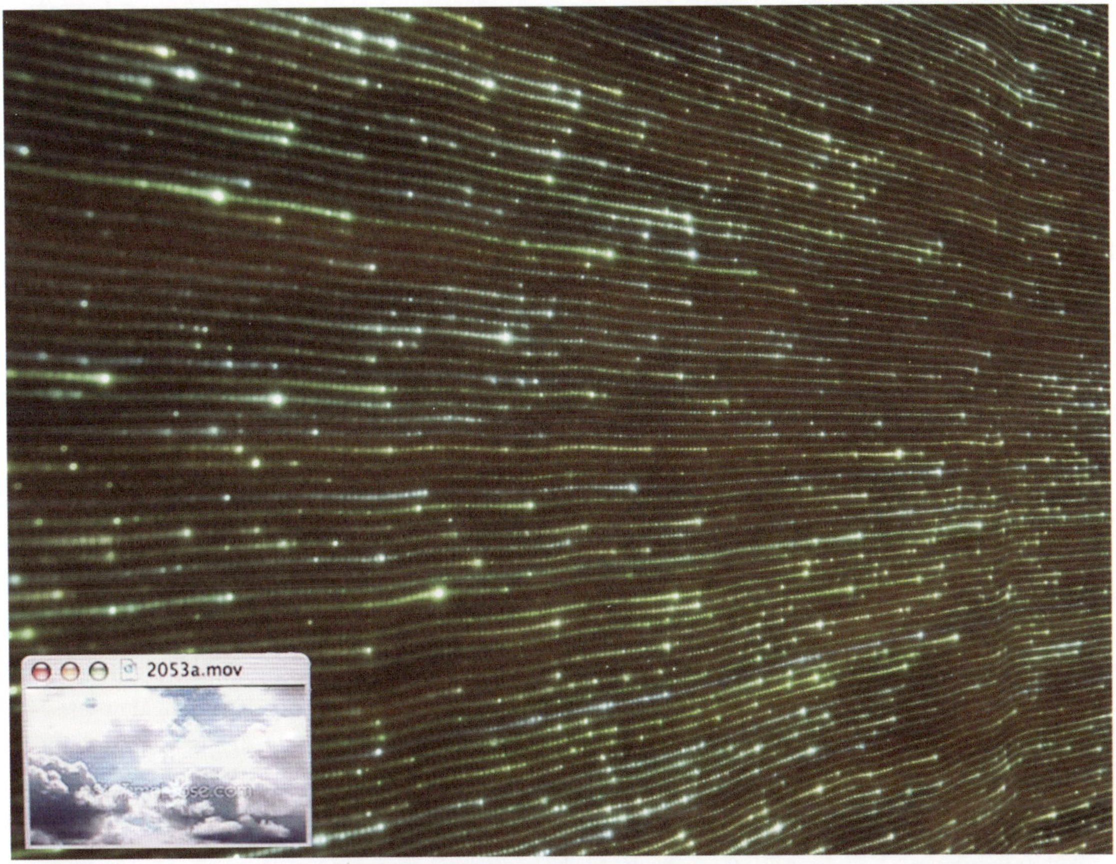
2053a.mov

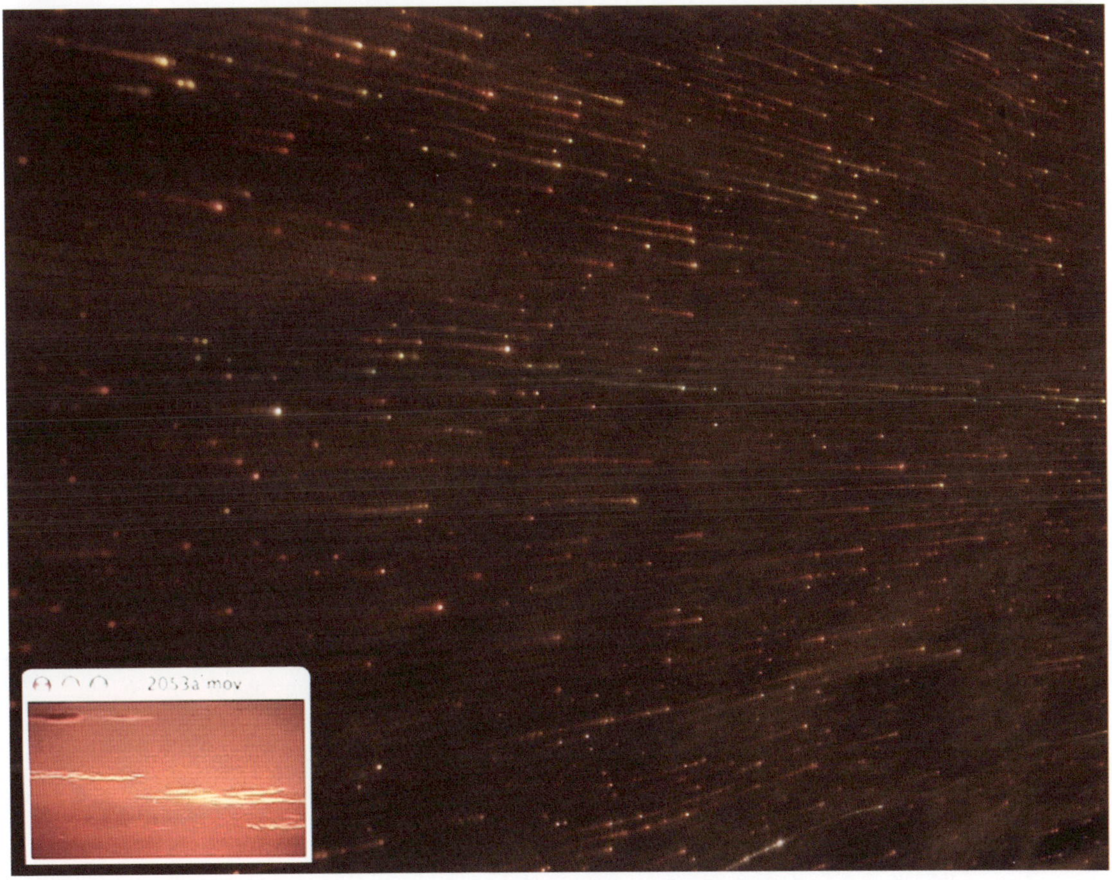
2053a.mov

engineering perhaps we may understand that the spatial boundary of an enclosure is not limited to the material surface but can be reconfigured as the zone in which change of energy fields occur. Thus smart materials enter the domain of architecture not as alternatives that replace existing static materials but as dynamic matter that alters their behaviour and capacities to respond to thermal, luminous, and acoustical energy fields.

Smart Fabrics

The field of high-performance textiles and flexible fibre-based materials is one of the most dynamic areas of material innovation currently reshaping the practice of fashion, industrial design, architecture, and engineering in a cross-disciplinary context. A 'fabric' refers to a material that in some way resembles or shares some of the properties of cloth either woven or flexible layered materials.[6] New fabrics as well as existing ones that integrate new material properties are progressively being tested in the field of construction to generate more transparent, lightweight, adaptable, and responsive environments that we can inhabit.[7]

Addington and Schodek also focus on actions and effects that are made possible via smart materials and technologies. Although many applications do exist exclusively in clothing, they point out that similar products and technologies that can be envisioned for use of smart fabrics in architecture and design. For example, if a coffee mug changes its colour based on the temperature of the beverage it contains, would it be possible to change the colour of a room based on its exposure to sunlight by integrating a photochromic textile on its walls? The authors organize high-performance and smart fabrics according to what they might do in a space.[8] The first group consists of the high-performance fabrics or flexible materials that are combined with other materials (composites or weaves) to accomplish some specific objective related to variables in the luminous, thermal, or acoustical environment and structural forces. These applications may lead to design of surfaces and structures that specifically reflect, absorb or transmit light, sound, and heat or react to building and gravitational forces in a designed way. The materials in this group, however, are not smart in the sense that they do not display changing characteristics, i.e. their properties are engineered for better performance yet remain static. The second group of fabrics exhibits some form of property change in response to the above-listed external conditions, most commonly colour-change based on impregnation or layering of thermochromic (heat-sensitive) or photochromic (light-sensitive) materials with the fabric. Fabrics in the third group provide an energy exchange function and are known mostly as phase-change materials. These fabrics involve absorbing, storing, or releasing large amounts of energy in the form of a latent heat and thus control thermal environments. For example, water is a phase-change material that transforms from solid to liquid to gas at freezing or boiling temperatures. This is not, however, of any use in construction industry as the energy exchange takes place at low or high temperatures.

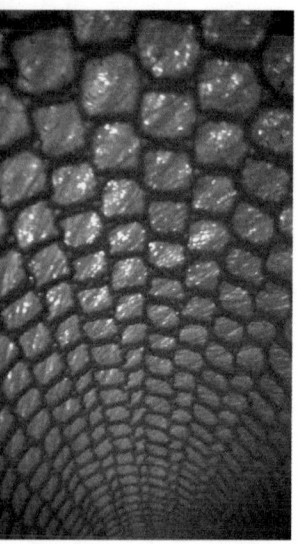

5　3-D Fabric

5　**D. M. Addington and D. L. Schodek**, *Smart Materials and New Technologies for the Architecture and Design Professions* (Oxford: Architectural Press, 2005), 7–8.

6　**Addington and Schodek**, *Smart Materials*, 158.

7　**F. Klassen** 'Material Innovations: Transparent, Lightweight and Malleable,' in eds. F. Klassen and R. Kronenburg, *Transportable Environments 3* (London: Taylor and Francis, 2006), 122–135.

8　**Addington and Schodek**, *Smart Materials*, 158–162.

Outlast Technologies Inc. has already developed a phase-change textile material for use in sports clothing (to keep users cool or warm). It hopes to develop a similar material for the building industry. Phase change particles can be encapsulated at microscopic levels and integrated into the fabric either as surface coating or an integral part of the fabric's fibre (by using a wet-spinning process).[9] And finally fabrics in the fourth category, known as electronic textiles, are in some way specifically intended to act as sensors and to be used in energy distribution or in communication networks.[10]

[9] **S. Braddock** *Techno Textiles: Revolutionary Fabrics for Fashion and Design* (New York: Thames and Hudson, 1999), 156.

[10] **Braddock,** *Techno Textiles,* 156–160.

Conclusion

These advancements in materials research exemplify the potential changes (functional, structural, design, and cultural) that may occur in the perceptual and practical characteristics of the built environments in the future. In return, continued demand for high-performance and responsive built environments will derive the specific material developments away from a mere experiment with the static qualities of materials towards multi-layered and active manipulators of external energy fields. The integration of building materials and the mechanics that regulate thermal, luminous, acoustical, and visual requirements is of vital importance in successfully translating and realizing innovative building concepts. We still are far from achieving a built environment that morphs itself for best performance in response to varying external forces. We are, however, slowly moving in a direction that is compatible with innovations that are reshaping other fields.

Notes

A longer version of this article was originally published in F. Klassen, 'Malleable Matter: Adaptable and Responsive Space,' *Proceedings of the Adaptables 2006 TU/E International Conference on Adaptability in Design and Construction, Volume Two* (Eindhoven: Technische Universiteit Eindhoven, 2006), 192–207.

Mobile Communication
and Education

Mobile Nation focuses on the value of art and design research in the field of mobile communication and related mobile technologies. Artists and designers use a range of approaches for creating new applications in this changing field. These include design charettes, participatory design, improvisation, information architecture and flow diagramming, mind mapping and modelling, engineering iteration, context-specific design, and user observation and interviews. *Mobile Nation's* roster of multi-disciplinary thinkers included accomplished researchers and scholars, technological innovators, industry experts, and graduate and undergraduate students who are expanding the field's knowledge of research, design, and engineering methodologies. In so doing, these players provide positive outcomes for the academic and educational, industrial, and design communities.

New questions concerning design research and education are emerging: In what context do education and learning methodologies figure in design research for mobile technology? How do we engage teens and young adults—our students are the largest demographic for mobile communication uptake and use—in this important research? In investigating the new forms of utility in educational environments, and the resulting evaluations of usability, the social nature of mobile use requires new and innovative research creation methods.

Gaming Literacy
game design as a model for literacy in the twenty-first century

Eric Zimmerman
Gamelab

Introduction: Literacy and games from the inside out

Gaming Literacy is an approach to literacy based on game design. My argument is that there is an emerging set of skills and competencies, a set of new ideas and practices that are going to be increasingly a part of what it means to be literate in the coming century. This essay's proposal is that game design is a paradigm for understanding what these literacy needs are and how they might be addressed. I look at three game design concepts—systems, play, and design—as key components of this new literacy.

Traditional ideas about literacy have centred on reading and writing—the ability to understand, exchange, and create meaning through text, speech, and other forms of language. A younger cousin to literacy studies, media literacy extended this thinking to diverse forms of media—from images and music to film, television, and advertising. The emphasis in media literacy as it evolved during the 1980s was an ideological critique of the hidden codes embedded in media. Media studies scholars ask certain kinds of questions: Is a given instance of media racist or sexist? Who is creating it and with what agenda? What kinds of intended and unintended messages and meanings does media contain?

Literacy, and even media literacy, is necessary but not sufficient for one to be fully literate in our world today. There are emerging needs for new kinds of literacy that are simply not being addressed, needs that arise in part from a growing use of computer and communication networks. Gaming literacy is one approach to addressing these new sorts of literacies that will become increasingly crucial for work, play, education, and citizenship in the coming century.

Gaming literacy reverses conventional ideas about what games are and how they function. A classical way of understanding games is the 'magic circle,' a concept that originates with Dutch historian and philosopher Johann Huizinga. The magic circle represents the idea that games take place within limits of time and space, that they are self-contained systems of meaning. A chess king, for example, is just a little figurine on a coffee table—but when a game of chess starts, it suddenly acquires all kinds of very specific strategic, psychological, and even narrative meanings. To consider another example, when a soccer game or *Street Fighter II* match begins, your friend suddenly becomes your opponent and bitter rival—at least for the duration of the game. While many social and cultural meanings certainly move into and out of any game (your game rivalry might ultimately change your friendship outside the game), the magic circle emphasizes meanings that are intrinsic and interior to games.

Gaming literacy turns this inward-looking focus inside out. Rather than addressing the meanings that only arise inside the magic circle of a game, it asks how games relate to the world outside the magic circle—how game playing and game design can be seen as models for learning and action in the real world. It asks, in other words, not 'What does gaming look like?' but instead 'What does the world look like from the point of view of gaming?'

But it's important to be very clear here: gaming literacy is not about just any kind of real-world impact—it is a specific form of literacy. So for the sake of specificity, here are some things that gaming literacy is *not*:

+ Gaming literacy is not about 'serious games'—games designed to teach you subject matter, such as eighth-grade algebra.

+ Gaming literacy is not about 'persuasive games' that are designed with a message or social agenda to impart to the player.

+ Gaming literacy is also not about training professional game designers, or even about the idea that anyone can be a game designer.

Gaming literacy is literacy—it is the ability to understand and create specific kinds of meanings. As I describe it here, gaming literacy is based on three concepts: systems, play, and design. All three are closely tied to game design, and each represents kinds of literacies that are not being addressed today through traditional education. Each concept also points to a new paradigm for what it will mean to become literate in the coming century. Together they stand for a new set of cognitive, creative, and social skills—a cluster of practices that I call *gaming literacy*.

I like the term gaming literacy not only because it references the way that games and game design are closely tied to the emerging literacies I identify, but also because of the mischievous double meaning of gaming, which can signify exploiting or taking clever advantage of something. Gaming a system means finding hidden shortcuts and cheats, and bending and modifying rules in order to move through the system more efficiently—perhaps to misbehave, but perhaps to change that system for the better. We can game the stock market, a university course registration process, or even just a flirtatious conversation. Gaming literacy, in other words, 'games' literacy, bending and breaking rules, playing with our notions of what literacy has been and can be.

Systems

To paraphrase contemporary communications theory, a system is a set of parts that interrelates to form a whole. Almost anything can be considered a system, from biological and physical systems to social and cultural systems.

Having a systems point of view, being systems literate, means understanding the world as dynamic sets of parts with complex, constantly changing interrelationships—seeing structures that underlie our world, and comprehending how these structures function.

As a key component of gaming literacy, systems can be considered a paradigm for literacy in the coming century. Increasingly, complex information systems are part of how we socialize and date, conduct business and do finance, learn and research, and conduct our working lives. Our world is increasingly defined by systems. Being able to successfully understand and navigate, modify, and design systems will be more and more linked to how we learn, work, play, and live as engaged world citizens.

Systems-based thinking is about process, not answers. It stresses the importance of dynamic relationships, not fixed facts. Getting to know a system requires understanding on several levels, from the fixed foundational structures of the system to its emergent, unpredictable patterns of behavior. Systems thinking thereby leads to kinds of improvisational critical problem-solving that will be key skills for creative learning and work in the future. In part, the rise of systems as an integral aspect of our lives is related to the increasing prominence of digital technology and networks. But systems literacy is not intrinsically related to computers. The key to systems literacy is about a shift in attitude, not about learning technological skills.

Systems literacy, as an approach to learning, is not wholly new. Theorists like Lev Vygotsky and approaches like constructivism and Montessori education have set some precedents. However, it goes without saying that a systems approach to literacy is the exact opposite of what is currently going on in American public schools under the standardized testing regime of the No Child Left Behind Act. Meanwhile, contemporary thinkers from Stephen Johnson to Malcolm Gladwell are increasingly proposing systems-based thinking as the best way to understand a range of complex subjects from media and society to history and culture.

If systems are a paradigm for an emerging form of literacy, what is the connection to games? Games are intrinsically, essentially systemic. Every game has a mathematical substratum, a set of rules that lies under its surface. Other kinds of media, art, and entertainment are not so intrinsically structured. Scholars debate, for example, the essential formal core of a film—Is it the script? The pattern of the editing over time? The composition of light and shadow in a frame? There is not one correct answer. But with games, there is the clarity of a formal system—the rules of the game. This formal system is the basis of the structures that constitute a game's systems. More than other kinds of culture and media which have been the focus of literacy in the past, then, games are uniquely well-suited to teach systems literacy.

To play, understand, and—especially—design games, one ends up having to understand them as systems. Any game is a kind of miniature artificial system, bounded and defined by the game rules that create the game's 'magic circle.' Playing a game well to see which strategies are more effective, analyzing the game's rules to see how they ramify into a player's experience, and designing a game by playtesting, modifying the rules, and playtesting again, are all examples of how games naturally and powerfully lend themselves to systems literacy.

Play

Games are systems in part because they are at some level mathematical systems of rules. But if games were just math, we would never have the athletic balletics of tennis, the bluffing warfare of poker, or the deep collaboration of World of Warcraft. Play is the human effect of rules set into motion, and play in its many forms play transcends the systems from which it arises. Just as games are more than their structures of rules, gaming literacy is more than the concept of systems. It is also play.

There is a curious relationship between rules and play. In the classical sense of a game as a magic circle, rules are fixed, rigid, and closed. They are logical, rational, and scientific. Rules really don't seem like much fun at all. But when rules are taken on and adopted by players, when they enter the magic circle and agree to follow the rules, play happens. Play in many ways is the opposite of rules: as much as rules are closed and fixed, play is improvisational and uncertain. Yet in a game, these two opposites find a common home—game play paradoxically occurring only because of game rules.

In Rules of Play, Katie Salen and I define play as 'free movement within a more rigid structure.' Imagine play as the 'free play' of a gear or steering wheel—the loose movement in an otherwise rigid structure of interlocking parts. The free play of a steering wheel is the distance it can move without engaging with the drive shaft, axle, and wheels—the more rigid utilitarian structures of the car. This free play only exists because of the more inflexible, functional structures of the automobile. Yet it also exists despite those structures, almost in spite of them. The play of a joke, for example, is funny because jokes play with structures of language, creating subtle ironies, or double-meanings, or vulgar inappropriateness. The free play humour of a joke exists in opposition to the more rigid structures of serious, ordinary language—yet is utterly dependent on these structures for its play.

Yet play is far more than just play within a structure. Play can play with structures. Players don't just play games, they mold them, engage in metaplay between games, and develop cultures around games. Games are not just about following rules, but also about breaking them, whether it is players creating homebrew rules for Monopoly, hacking into their favorite death-match title, or breaking social norms in classics like spin the bottle that celebrate and create taboo behavior.

Although play exists outside of games, games do provide one of the very best platforms for understanding play—from free play within a structure to the transformative play that reconfigures that structure. Any instance of a game is an engine designed to produce play, a miniature laboratory for studying play qua play.

So why is play an important paradigm for literacy in the coming century? Systems are important, but if we limit literacy to structural, systemic literacy, then we are missing part of the equation. When we move from systems to play, we shift focus from the game to the players, from structures of rules to structures of human interaction. Games as play are social ecosystems and personal experience —and these dimensions are key aspects of a well-rounded literacy.

As our lives become more networked, people are, more and more, engaging with structures, but they are not just inhabiting them. They are playing with them. A social network like Wikipedia is not just a fixed construct like a circuit diagram. It is a fuzzy system, a dynamic system, a social system, a cultural system. Systems only become meaningful as they are inhabited, explored, and manipulated by people. In the coming century, what will be important is not just systems, but human systems.

A literacy based on play is a literacy of innovation and invention. Just as systems literacy is about engendering a systems-based attitude, being literate in play means being playful – having a ludic attitude that sees the world's structures as opportunities for playful engagement. What does it mean to play with institutional language, with social spaces, or with processes of learning? When these rules are bent, broken, and transformed, what new structures will arise?

Play emerges from more rigid systems, but it does not take those systems for granted. It plays with them, modifying, transgressing, and reinventing. We must learn to approach problem-solving with a spirit of playfulness, to not resist but in fact embrace transformation and change. As a paradigm for innovation in the coming century, play will increasingly inform how we learn, work, and create culture.

Design

The notion of design connects powerfully to the sort of creative intelligence the best practitioners need in order to be able, continually, to redesign their activities in the very act of practice. It connects as well to the idea that learning and productivity are the results of the designs (the structures) of complex systems of people, environments, technology, beliefs, and texts.
— *The New London Group,*
"A Pedagogy of Multiliteracies: Designing Social Futures"

If gaming literacy were simply about systems and play, it would be a literacy based on games, not game design. But design, the third component of gaming literacy, is absolutely key, and in many ways brings the traditional idea of literacy as understanding and creating meaning back into the mix. There are many definitions of design, but in *Rules of Play* Katie Salen and I describe design as 'the process by which a designer creates a context, to be encountered by a participant, from which meaning emerges.'

Design as the creation of meaning invokes the magic circle: designers create contexts that in turn create signification. Although design comes in many forms, from architecture to industrial design, games happen to be incredibly well-suited for studying how meaning is made. Outside the game of rock paper scissors, a fist can mean many things. But inside the game, that gesture is assigned a highly specific significance, a defined meaning within the lexicon of the game's language. The creation of meaning through game design is wonderfully complex. A game creates its own meanings (blue means enemy; yellow means power-up), but also traffics with meanings from the outside (horror film music in a shooter means danger is coming; poker means a fun evening with friends).

For a game designer, the creation of meaning is a second-order problem. The game designer creates structures of rules directly, but only indirectly creates the experience of play when the rules are enacted by players. As a game unfolds through play, metaplay, and transformative play, unexpected things happen, patterns that are impossible to completely predict. In this way, design is not about the creation of a fixed object. It is about creating a set of possibilities. The audience is always at least one step removed from the designer. Games embody this aspect of design in a very direct and essential way; even the most straightforward game of chess or *Sims* is about players exploring possibilities they are given by a designed object. In a game, design mediates between structure and play; a game system is designed just so play will occur.

Over and above game design's affinity for the process of making meaning, it is also radically interdisciplinary. Making a game includes creating a formal system of rules, while also designing a human play experience for a particular cultural and social context. Game design involves math and logic, aesthetics and storytelling, writing and communication, visual and audio design, human psychology and behaviour, and understanding culture through art, entertainment, and popular media. For video game design, computer and technological literacy become part of the equation as well.

As an exploration of process, as the rigorous creation of meaning, and as a uniquely interdisciplinary endeavor, game design represents multimodal forms of learning that educators and literacy theorists have been talking about for years, perhaps most significantly in the publications of the New London Group. Game design, as the investigation of the possibility of meaning, truly gets at the heart of gaming literacy and ties together systems, play, and design into a unified and integrated process.

Conclusion: A playful world

As we move into the early years of the twenty-first century, the world is becoming increasingly transformed by communications, transportation, and information technology that is shrinking our globe, making it a place of cultural exchanges both constructive and destructive. Existing models of literacy simply do not fully address reality in the world today.

Gaming literacy is certainly not the only way to understand the emerging literacy needs I have identified. But games and game design are one promising approach, making use of a cultural form that is wildly popular and wildly varied, both incredibly ancient and strikingly contemporary. And intrinsically playful as well.

So how does one take action to promote gaming literacy? At Gamelab, the independent game development company I founded in 2000 with Peter Lee, we have begun a number of gaming literacy projects. We are building *Gamestar Mechanic*, funded by the MacArthur Foundation and created in collaboration with the Games and Professional Practice Situations (GAPPS) group at the University of Wisconsin-Madison—a computer program that will let youth learn about game design by letting them create and modify simple games. In June 2007 Gamelab announced creation of the Gamelab Institute of Play. With Katie Salen as its executive director, the institute will promote gaming literacy through educational programs and advocacy.

What does gaming literacy mean for game players and game makers? The good news is that games, so often maligned, have much to offer our complex world. And not just so-called "serious games" with explicit educational goals, but *any* game. Gaming literacy can help us feel good about what we do by playing games, making games, studying games, modifying games, and taking part in gaming communities. As literacy scholar James Gee likes to say, "video games are good for your soul."

Gaming literacy turns the tables on the usual way we regard games. Rather than focusing on what happens inside the artificial world of a game, gaming literacy asks how playing, understanding, and designing games embodies a crucial way of looking at and being in the world. That way of being embraces the rigour of systems, the creativity of play, and the game design instinct to continually redesign and reinvent meaning.

It's not that games will necessarily make the world a better place. But in the coming century, the way we live and learn, work and relax, and communicate and create is going to more and more resemble how we play games. And while we're not all going to be game designers, game design and gaming literacy offers a valuable model for what it will mean to become literate, educated, and successful in this playful world.

Coda: No essay is an island

The ideas in this essay are not just my own, but are part of a growing conversation that can be heard across universities, commercial game companies, grade-school classrooms, non-profit foundations, and in other places where game players, game makers, scholars, and educators intersect.

Although I have been a game designer and game design theorist for more than a decade, I began to rigorously connect game design and literacy through my interaction with the GAPPS group (now called Games, Learning, and Society, or GLS), a collection of scholars at the University of Wisconsin-Madison that includes Jim Gee, Rich Halverson, Betty Hayes, David Shaffer, Kurt Squire, and Constance Steinkuehler. I was privileged to be invited to a series of conversations with this stimulating crew about games and literacy sponsored by the Spencer Foundation. In 2006, during the third of these three meetings, the term 'gaming literacy' emerged from our conversations as a concept that could reference growing connections between games, learning, literacy, and design.

I am greatly indebted to game designer, scholar, and educator Katie Salen for our ongoing collaborations, including the textbook *Rules of Play: Game Design Fundamentals*. My ideas on game design and learning have also been shaped by my work with the amazing staff at Gamelab, especially my co-founder Peter Lee, and former Gamelab game designers Frank Lantz and Nick Fortugno. Connie Yowell at the MacArthur Foundation also has been instrumental in bringing together scholars, artists, educators, and designers to exchange ideas, including commission of important foundational research by media scholar Henry Jenkins. The specific formulations in this paper were first instantiated in a talk I gave at Vancouver's Simon Frasier University in January 2007, and this essay received valuable feedback from Jim Gee, Katie Salen, Kurt Squire, and Constance Steinkuehler.

I go to this trouble to highlight some sources in order to emphasize the newness of these ideas and the collaborative way that they are emerging from a thick soup of scholarship, debate, and collaboration. This kind of dialogue is very much in the spirit of gaming literacy itself, and I encourage others to take part in the conversation as well. Some of the best places to get involved include the GLS conference held annually at the University of Wisconsin-Madison (http://www.glsconference.org) , the Serious Games Initiative (http://www.seriousgames.org), the Education Special Interest Group of the International Game Developers Association (http://www.igda.org/education), and the ongoing dialogues about digital media literacy at the MacArthur Foundation website (http://community.macfound.org/openforum).

Notes

This essay originally appeared in the premier issue of the *Harvard Interactive Media Review*, and is reprinted in its original text, without concession to standard Canadian spellings or grammatical forms, by request and kind permission of the author. Its inclusion acknowledges the important and lively contribution to the discussions on design methodologies and gaming, as applied to mobile technologies, made by Eric Zimmerman at the Mobile Nation conference and to this book.

Mobile Sphere-ing
methods for making virtual spaces public

Paula Gardner
Ontario College of Art & Design

There has been much debate amongst critical theorists over the years regarding the possibility of a so-called public sphere in democratic capitalist societies—a space allowing citizens to insert themselves into narratives of public culture and civic responsibility as agents of community and nation building. The public sphere is often imagined as a collection of private individuals seeking a common good, where industry is deliberately excluded. Where Jurgen Habermas famously argued that the (nineteenth-century) public sphere had been destroyed by consumer culture, facilitating a rise of media and state intervention in the 'private,'[1] recent critiques have contended that globalized industries have elided state sovereignty and prohibited public formations. A coherent public sphere is a nonsensical concept in many ways, assuming preconstituted, homogenous, and coherent populations separate and juxtaposed to both the state and an (allegedly) distinct private sphere. Many critical theorists have not, however, given up the fight to designate some space of action for individuals who are socially, economically, and politically marginalized by globalization practices and policies. A reimagination of this sphere must, then, take into account the various impacts of transnational flow on the communication possibilities for 'publics' of varying scales: local, national, diasporic, transnational, etc.

This paper considers how a radicalized public space might be facilitated via mobile technologies and experiences and the methods by which they are designed. Ernesto LaClau and Chantal Mouffe, for example, imagine 'agonistic' democracies that require agitation and debate and create channels that facilitate different values; such processes demand the 'democratization of democracy,'[2] holding true to values routinely espoused but violated by states.[3] There are resonances of this rigorous democratic practice as well in Nancy Fraser's call for justice that crosses national boundaries, as it must in this age of transnational flow. In the resulting global public sphere, states, industries, and international structures are all held accountable for injustices. To this end, Fraser argues we must insist that states address the 'how' by which structural and institutional frameworks of societies set ground rules that govern social interaction.[4] I want to think about how different kinds of publics (i.e. diverse citizens, academics, industry workers, etc.) in a variety of scaled public geographies (global, local, national, and other) can take part in such practices to democratize and hold the state accountable for broadly framed notions of justice.

facing page
1 **PORTAGE** visualization for John Street, Toronto

2 **Alter Audio charette**
Students were challenged
to create innovative audio
experiences

I want to think as well about everyday practices of communication that become colonized by industrial and cultural trends of 'framesetting,'[5] such as those pervading the Internet, and consider how greater democratic processes might be achieved via social interactions made possible via mobile technologies.[6]

This paper begins to analyze how we can engage with new communication technologies at the levels of design, production, and use, from a stance committed to agonistic democratic processes that open up channels for debate and ultimately 'reframe' our value systems, making them accountable to radical democratic roots. What roles do technology, game/experience designers, and users of mobile media play in reframing these processes and places of communication that are, after all, essential to pluralistic democracy? Might it be possible that the spatial dimensions of mobile technologies, and for that matter the intersections of space with time therein, differ from the Internet experience, precisely and most obviously because the mobile subject can move through material space utilizing technologies that themselves move through a variety of space and time dimensions? Where the Internet allows users to negotiate virtual space, mobile technologies facilitate user interventions in both spheres, at will and even simultaneously, as we negotiate diverse and fluid temporal and spatial dimensions of experience. Might we, then, analyze mobile technologies—both the spaces of content and architecture and the mobile spaces in which they operate—as places for public culture to represent, identify, present, agitate, create, craft movements, and, as well, demand new structures and forums for justice?

The highly unique spaces of mobile technologies, made possible by the Hertzian/virtual/material environs they traverse, uniquely position users as unfolding and changing producer-consumers. Where a differentiated and evolving subjectivity is *possible* on the Internet, mobile spaces engage subjects in an ongoing negotiation of and play with the false dichotomies of public/private; inside/outside; narrowcast/broadcast; active/ passive; linear/non-linear; self/other; material/virtual; analog/digital; art/design; empiricism/creation; etc. Because mobile technologies and their users negotiate multiple spaces simultaneously, they allow for play that interrogates these dualistic schisms and discomforts in distinctive ways. Lev Manovich, for example, imagines mobile media as enabling a 'poetics of augmented space';[7] the mobile subject moves in the lived experience and aesthetics of augmented space—'physical space plus layers of (mobile) data'—where unique subjectivity and art can be created. This subject is simultaneously material and postmodern, fluidly crossing boundaries, engaging interactively and alone—a subject in progress negotiating experiences that are temporally and spatially varied.[8] This hybrid, incoherent subjectivity seeks engagement over progress, experience over end, favours no actor over another, and, to that end, is entirely in keeping with subjects of agonistic democracy in encouraging diverse interactions.[9] The hybrid, porous spaces of mobile technologies, then, might facilitate a messy kind of radical democracy or some semblance of a postmodern public sphere.[10]

MOBILE COMMUNICATION AND EDUCATION

These augmented spaces offer opportunities to designers of mobile games and experiences, as well as hardware and software designers, to consider user experiences that feed agonistic public practices. As an academic and media producer participating in mobile experience design, I wish to think through the possibilities of working on experience and software design from this standpoint.[11] Two mobile design projects I have worked on have yielded both intentional and unintentional methodologies and practices that have potential utility for putting the 'democratic process back in democracy.' Each project has been structured to encourage intensive collaborative among a diverse team of designers, engineers, and users,[12] and has relied on methods of critical ethnography and what can be termed 'rapid iterative prototyping.' In our recent project, entitled PORTAGE: The Canadian Mobile Experience,[13] we are employing these methods in partnership with small- and medium-sized industries to create, collectively, a virtual mobile theatre in downtown Toronto .

Our methodology is driven by reflexive brainstorming practices that call for the ongoing renegotiation of our methods and goals until we reach consensus. One of our team's few rules is to continually imagine the intended user experience, while prioritizing possibilities for social interaction and authoring. This process, which was for us surprisingly functional and efficient, contests neo-liberal mandates of cultural practice; it continually destabilizes any position of authority or privileged knowledge, and arrests the formation of epistemological monopolies or their fruition in design strategies. In embedding self-reflexivity, the model rejects any notion that (democratic) social interactions can be universally facilitated—instead mobile designs should speak specifically to both the time and space in which the interactions occur.

Like most experimentation, our design teams discovered serendipitous, often accidental, innovative processes. The Alter Audio team (of the Mobile Digital Commons Network, or MDCN), for example, aimed to exploit the aural bias of the mobile phone to create unique collaborative experiences. Our design foregrounded user choice, allowing 'players' to decide whether to combine musical sounds with another, and making it possible for users to sample or upload and trade sounds in musical collaboration. The privileging of the user also grounded our commitment to an iterative design strategy that called for the analysis, design, construction, and implementation of *complete* experiences, and a cycling through this process until the experience was satisfying.[14] Through these practices, we discovered the benefits of forging rapid prototyping to iterative design. This melding reified the value of working within limits in order to push them and thus reimagine and alter the frame.

The best example of this process comes from charettes held by Alter Audio, whereby three Ontario College of Art & Design (OCAD) undergraduate students were challenged to create, in two short days, innovative audio experiences within highly restrictive iterations designed for a mobile phone.[15] The students were given minimal guidance regarding the two iterations available to them, and little information about the goals of Alter Audio or MDCN. Their specific challenge was, over the course of three

3 Proximity
Sounds are controlled by the user. Via Bluetooth, sounds are also triggered by the proximity of other users.

days, to employ one of the two design architectures, using Bluetooth or GPS, to trigger sounds (housed in the phones), and to create and deliver a unique audio experience for two to four individuals.[16] The students rallied against these limitations, imagined more complex iterations, and finally began brainstorming and designing experiences. Each student proclaimed different design aims, respectively: to enhance the environment (with locative-signified symphonies), to build community, and to evolve content that could be employed in future, more complex, iterations. Interestingly, each student's design suggested that mobile experiences are distinctive in involving the movement of bodies through augmented spaces; as such their designs encouraged, and sometimes intentionally solicited, new social interactions.

One student created an experience named *Vocal Chords*, employing the human voice as his medium, and taking advantage of iteration two's inability to sync sounds; he recorded a male and female voice singing notes in the key of C that combined interminably, and thus created strange and unexpected chords that were triggered by Bluetooth proximity to another user. Another student, also exploiting this design limitation, composed *Jam*, an audio choreography combining diverse sounds including five-second loops of long, sustained tones, nature (birds), and syncopated percussion in a triplet rhythm; due to their disparate textures and tempos, the sounds combined equally well, despite the asynchronic bias.

Neither student designer would have created these content or experiences if not pressed by the rapid iterative design method. These results were significant in that each student imagined the iteration's constrained frameworks as opportunities for massaging, pressing, and reimagining the frame. This manner of conscious engagement with technologies and their limits demonstrates the best of agonistic practices that work to maintain structures even while they disband the (dominant) logics associated with them. These acts of internal resistance were made possible by the paradigm of our design strategy, including the (intentional) minimal context team leaders provided to designers to qualify the frames/iterations, our insistence that experiences be delivered in short time, and an overarching iterative methodology that prioritized user choice and experimentation. The resulting designs disrupted the false dichotomy of open and closed systems that assumes open systems alone allow for the reinvention of rules and feed innovation.[17] Here, a hybrid of open and closed processes created unforeseen designs and user experiences.

Our focus on user desire is also privileged through testing completed iterations on a range of users. We employ critical ethnographic methods that seek to 'purge' users of information regarding the mobile experience in an open-ended interview or focus group format. These methods allow users to experience various mobile experiences in the field while our team observes, and carefully avoid suggesting to users the goals or intent of the experience to be tested. In addition, we collect information that contextualizes the users' responses to the experiences; these include users' media consumption and purchasing habits, functional and social habits of mobile and digital technology use, and favoured communication strategies, as well as cultural practices and backgrounds.[18] After researchers contextualize user comments in relation to this data, users are asked to respond to 'findings' to ensure they accurately reflect their readings of the experience. Vetted findings are then cycled back into the next design iteration. This type of critical ethnographic methodology shares assumptions with agonistic democracy in challenging narrow, often cloistered practices of research and design, overthrowing the expert/user dichotomy, and ensuring—indeed seeking—an array of voices that might otherwise be marginalized or ignored in the design process.

Because access to technology is a challenge for both mobile users and our team, this problematic also plays a key role in our design strategies. Access to mobile technologies and network services in North America is routinely constricted due to a few companies monopolizing the industry. In turn, this phenomenon tends to embed neo-liberal imperatives (i.e. of commercialization, consumption, and limited democracy) in design structures and experiences. For example, mobile industries tend to design for the 'masses,' with such experiences as competitive games with commodity rewards, copying pre-existing games that possess mass-market appeal (e.g. *Tetris* or Monopoly), or employing the forum as the next digital 'suburbia,' to circulate (or even spam) consumers with commercial promotions. But rarely do mobile designs prioritize social interaction, as is the tendency of artists and independent designers who work on mobile platforms.

Because it is designed to bridge the digital divide in a public, urban setting, PORTAGE in particular has been forced to contend with the limited features enabled on consumer-grade mobile phones.[19] Designers on similar projects, who seek to create broad access to mobile experiences with ease of use and low cost, and who refrain from hacking systems, are thus forced to dialogue with cellphone companies for access or to brainstorm new forms of connectivity.[20] The PORTAGE project, for example, is producing a paper critiquing the impact of industry monopolies upon mobile design research, and is dialoguing with service providers to discuss the resulting injury to both users and innovation, even as we experiment with alternatives to connecting via cell networks.

Importantly, however, PORTAGE has also discovered unexpected common ground with the innovators of small industry. When PORTAGE has placed issues of access, diversity of content, social interaction, and the public sphere on the table for discussion with industry partners,[20] we have moved together toward a mobile commons, rather than toward neo-industrial motifs or ideals. One of PORTAGE's most radical methods, in fact, has been the intensive collaboration among our team's small industry representatives, artists, designers, engineers, social scientists, and academics. As well, the team has, as a whole, committed to creating opportunities for users owning no mobile devices; such users will be able to engage instead through analog experiences that trigger a digital response or effect and thus allows social or technologically convergent interactivity for all passersby.[21]

The negotiations with industry leaders have been fruitful and sur-prising—we have together vetted our assumptions, interests in outcomes, and are melding a hybrid PORTAGE sensibility. To our surprise, we have realized the following *common* concerns: bridging the digital divide via mobile and hybrid technologies, creating aesthetic experiences that are grounded in theoretical questioning of linear history-making and public surveillance practices, allowing designers to lead design, and prioritizing social interaction and user choice. Perhaps this collision is due to the histo-ric phenomena whereby small- and medium-sized industries (in North America) have acted as media and cultural content innovators, rather than bending for the lowest common denominator consumer (as does big busi-ness). As well, smaller scale design demands greater democracy to preserve a productive work community. And yet the industry partners' interest in bringing art, social interaction, Canadian heritage, and political and social issues to mobile experiences, rather than selling their existing cultural content products is remarkable, and their enthusiasm is palpable. This collaborative realization is a step toward design models that build hybrid, activated publics.

4 Metaphors and Data Caches
The environment affects the musical composition through collaborative or choreographic options and data caches

Conclusion

Rapid, iterative prototyping can work to establish designs for interaction consistent with agonistic space and those required in a global digital commons. Fraser, LaClau, and Mouffe call not for the dissolution of existing structures but the need to re-evaluate structural frames to allow for resistive, alternative, and agonistic participation. Our integration of student designers and users serves as a case studies in democratized design practice, in allowing users with no particular (professional) quali-fications to determine the level of social engagement they desire in the very *paradigm* of the experience, and in ensuring that user experience remains the focal point of experience design. In PORTAGE, Alter Audio, and other projects, a decentralized, hybrid design team employing self-reflexive methods has yielded projects that in many ways reflect a postmodern, post-global ideological bias. We have produced cultural content, experi-ences, and ideas that, like Jean Baudrillard's 'simulations,' don't reflect any existing experience or seek to mimic existing games, but have emerged from a hybrid model, that is both open and closed, self-reflexive, and rigorously public.[22]

Mobile spaces, as such, offer possibilities for an ongoing negotiation of subjectivity in terms of time, space, and aesthetic, a reinterpretation of the nature of social interaction in the global technology age, and the forging of new relationships evolving self, society, and technology. Distinctively, mobile users cross borders of space and time as they physically and virtually traverse augmented spaces that are, as such, hybrid and political. Perhaps, then, mobile spaces and experiments are primed for a variety of publics to query the possibility of a global digital commons.

Notes

1 Jurgen Habermas, *The Structural Transformation of the Public Sphere*, trans. Thomas Burger (Cambridge: MIT Press, 1989).

2 Ernesto LaClau and Chantal Mouffe. 'Post-Marxism without Apologies,' *New Left Review* I/166 (Nov-Dec 1987); Dave Castle, 'Hearts, Minds and Radical Democracy: An Interview with Ernesto LaClau and Chantal Mouffe,' *Redpepper*, http://redpepper.org.uk.

3 Unfortunately, the space available precludes a deeper analysis of the complex theoretical debates regarding the meaning of 'space.'

4 Nancy Fraser, 'Reframing Justice in a Globalizing World,' *New Left Review* 36 no. 6 (Nov-Dec 2005); Nancy Fraser 'Transnationalizing the Public Sphere,' *RePublic Art*, www.republicart.net/disc/publicum/fraser01_en.htm.

5 Fraser, 'Reframing Justice,' 2-3, 10. This is Fraser's term to describe the 'Keynesian-Westphalian' dominant framework by which western states create structures of justice; it refers to the 'national-territorial underpinnings of justice dispute' occurring during the postwar democratic welfare state period of 1945 to 1970. The system 'mapped the world as a system of mutually recognizing sovereign territorial states.'

6 Mark Poster, for example, argues that the Internet 'is above all a decentralized communication system,' although he rejects determinist suggestions that it institutes radical decentralization or can decentralize social interactions. Conversely Greg Elmer has argued that the architecture of the Internet, because it is largely governed by industry monopolies that control search engines, requires consumers to allow companies to survey their Internet travels. In turn, companies repackage this data in topographical maps of consumers that result in a variety of marketing campaigns. Consumers and information seekers, then, become profiled by industry (and certainly, governments who have access to such data) and become participants in their own subjection/profiling. See Mark Poster, 'Cyberdemocracy: Internet and the Public Sphere,' originally published 1995, archived at www.humanities.uci.edu/mposter/writings/democ.html; Greg Elmer, *Profiling Machines: Mapping the Personal Information Economy* (Cambridge: MIT Press, 2004).

7 Because electronically augmented spaces personalize information for users dynamically change over time, and are delivered interactively, they bring up design issues that are not only technological, but conceptual. In other words, the problem of 'overlaying dynamic and contextual data in a physical space is a particular case of a general aesthetic paradigm: how to combine different spaces together.' The architecture of augmented space is layered with context, offering both challenge and opportunity at the technological and conceptual levels. Lev Manovich, 'The Poetics of Augmented Space' in eds. Anna Everett and John T. Caldwell, *New Media: Theories and Practices of Digitextuality* (New York: Routledge, 2003), 75–92.

8 Like Nina Wakeford's real-time video of a bike messenger (presented at Mobile Nation, with a related paper included in this volume), she never gets where she is going. Getting there is not the point.

9 As Drew Hemment's Location Oriented Critical Arts (LOCA) project demonstrates, mobile technologies are an excellent venue for interrogating the surveillance culture in which we are embedded. The project can be viewed at http://leoalmanac.org/gallery/locative/loca/index.htm.

¹⁰ Rob Shields, in his presentation and paper for Mobile Nation (also included in this volume), argues that the phone both links and binds, representing a 'coming into relationship with' (via photography). More, there is something 'unique' about the time and space elements of the mobile experience that operates differently from the Internet, the performance of mobile photography, for example, representing a 'virtual and present' melding.

¹¹ Designers also carry the extra currency of influencing engineers—in our case, designers work in close collaboration with engineers who have in turn pushed software and will likely create future mobile hardware development.

¹² Our diverse team consisted of a range of academics, artists, software designers, engineers, student artists, designers, and random (unconnected) users. Brainstorms that were rigorously open resulted in consensus, and the ultimate project design could be claimed by no single team member. I have worked as a researcher on MDCN, funded by the Canadian Culture On-Line Program of Canadian Heritage, and as a co-principal investigator, currently, on PORTAGE: The Canadian Mobile Experience, funded by the New Media Research and Development Initiative of Canadian Heritage.

¹³ The co-leads for this project, funded by the New Media Research and Development fund of Canadian Heritage, are myself and Geoffrey Shea. The project can be followed at www.ocad.ca/portage. As one walks down John Street, the heart of Toronto's art and entertainment district, viewers will be able to interact with a variety of art and entertainment content (industry and artist-created) to deposit, upload (audio/music jams, video, etc.), to create virtual graffiti, or conduct counter-surveillance of one's self on a mobile device.

¹⁴ This method is, of course, in contrast to segmented design processes that allow for the production of coherent (perhaps artistic and beautiful) elements that might seem to enhance an experience, but in fact often work better in isolation because they were designed that way.

¹⁵ Many projects within the MDCN utilized these experimental design techniques. The term 'charette' comes from nineteenth-century practices in France at the École des Beaux Arts, where architecture students were challenged to solve design problems in a distinct time period. As such, students rushed their drawings to the school from their studios in donkey carts or 'charettes,' hence the term.

¹⁶ These iterations and the charette methods are further outlined in Paula Gardner and Geoffrey Shea, 'Alter Audio: Mobile and Locative Sound Experiences,' Mobile Digital Commons Network website, www.mobilelab.ca/alteraudio. Iteration two supplied four audio tracks that turn on and off in different combinations and loop with no particular synchronization when users move in and out of Bluetooth range. Bluetooth range during our testing was somewhere between three to ten metres depending on the electronic 'noisiness' of the environment, the number of phones involved, physical structures like walls, etc. By design, a single user could not effect a particular combination of sounds by themselves. Iteration three involved the same constraints, but sounds were triggered by GPS nodes.

¹⁷ Kim Sawchuck, in her paper at Mobile Nation, published in this volume, warned of assuming that open systems are necessarily juxtaposed to closed systems.

¹⁸ These methods are in the same vein as those used by cultural anthropologists Sherry Turkle and Joseph Dumit. See Sherry Turkle, *The Second Self: The Human Spirit in the Computer Culture* (New York: Simon and Schuster, 1984); and Joseph Dumit, *Picturing Personhood: Brainscans and Biomedical Identity* (Princeton: Princeton University Press, 2003).

19 Cellphone features such as Bluetooth are disabled by network carriers in North America, who then sell users access to these services. In the US, a few large cellphone companies—Verizon and AT&T—won't allow features on their phones if their networks can't control them. As blogger Jan Frel comments, this constrains the 'mobile economy.' In addition, and causing equal damage, it constrains the formation of a global public sphere. Frel also notes that the Federal Communications Commission is about to auction off spectrum, creating the opportunity for a wireless broadband wholesaler to buy up and rent their network to a company that links cellphone and broadband; this could ultimately allow cellphone access computers that yield broadband to users. Jan Frel, 'Why You Can't Get Your iPhone,' *AlterNet*, posted April 13, 2007, www.alternet.org/bloggers/frel/50561/

20 Our industry partners include ecentricarts, Bravo!FACT, Collideascope Digital Productions, Triptych Media, and DECODE Entertainment, who produce, respectively, interface design, mobile film, integrated TV and new media, film and television animation, and family animation. In addition, our presentation partners are the Design Exchange, a design research and exhibition centre, and InterAccess, an art/technology art centre.

21 These might, for example, be hybrid analog-digital devices that individuals can kick or hit (e.g. a musical instrument) to create digital effects, while others play with mobile devices.

22 Jean Baudrillard, 'Simulacra and Simulations,' from *Jean Baudrillard, Selected Writings*, ed. Mark Poster (Stanford: Stanford University Press, 1988), 166–184.

Inside-out Experience Design

Geoffrey Shea
Ontario College of Art & Design

This paper looks at one artistic methodology or process for designing objects, images, and systems for a new context, in this case the emerging medium of mobile devices. In the research referenced here, mobile devices consist primarily of cellphones with ancillary components such as GPS locators and environmentally installed beacons such as Bluetooth transponders. Although cellphones cannot really be characterized as 'emerging,' their transition from a personal communication device to a mass-media vehicle is in its early stages.

The methodology outlined here was first adopted by the Alter Audio research team led by Geoffrey Shea and Paula Gardner as part of the Mobile Digital Commons Network (MDCN), a multi-year project undertaken by Concordia University, the Banff New Media Institute, and the Ontario College of Art & Design.[1] It is being further refined in PORTAGE, a research and development project led by Gardner and Shea at the Ontario College of Art & Design.[2]

Inside-out Experience Design

Perhaps because of the inherent two-way communication potential of cellphones and related devices, one principle often adopted in developing for this platform involves dynamic participation by the user. Participation in these examples includes active consumption, collaboration, co-creation, and user input, and mirrors developments in other media such as remixing, blog publishing, online videocasting, and immersive world games.

At the core of the artistic and research methodology employed in these projects is the principle of the iterative design cycle. In contrast to other, mainly software design, strategies, the iterative approach assumes that creators will make similar things over and over, improving or extending them with each new case. (Top-down design starts with a definition of the final product and then develops components and sub-components as needed. Bottom-up design examines existing capabilities and then seeks to find applications for them. Participatory design assumes there is already a large user base that can be engaged in the refinement process.[3] And extreme programming, the closest cousin to the iterative model, relies on fast implementation to counteract the inevitable evolution of the initial project requirements.)[4]

1 **Shea directing** research assistants during a charette

2 **Audience member** Randy Martin listening to *Phone Noir* during the Words Aloud! poetry festival.
John Pavicic working during a charette.

facing page
3–4 **PORTAGE concept map** prepared by Ken Leung following a number of team brainstorming sessions. OCAD concept map prepared by Patricio Davila following MDCN brainstorming sessions.

portage concept map
graphics
video output/ input
mobile device input
communication tech
social/ narrative
techniques
concepts
locations
landmarks
environments
bureacracy

Treatment
User Scenario
Diagram
Visual Images
Audio transmission game - caches on mountain - people would record
Samples + exchange - cultures from communities; remote is on mountain
Narrative - ghosts - users on path - hear voices; p.o.v. stories, Aboriginal e.g. follow that guide
Democratic Audio
Haunting
How get 2 play?
How fun is this?
Where hidden?

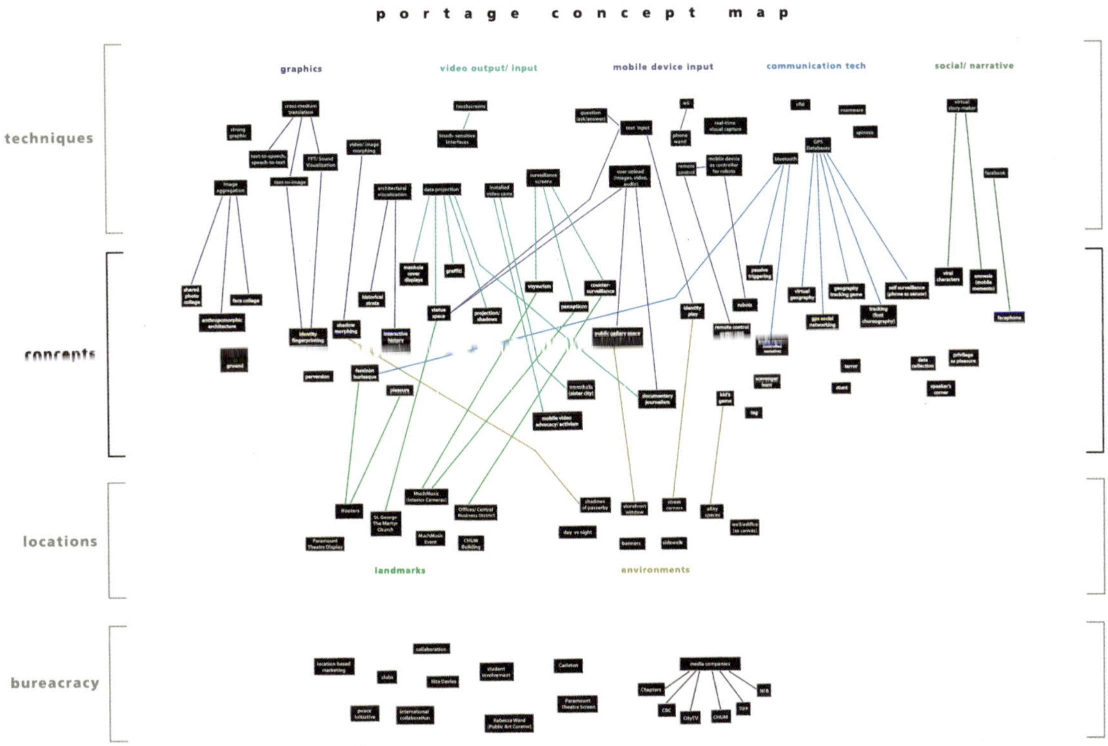

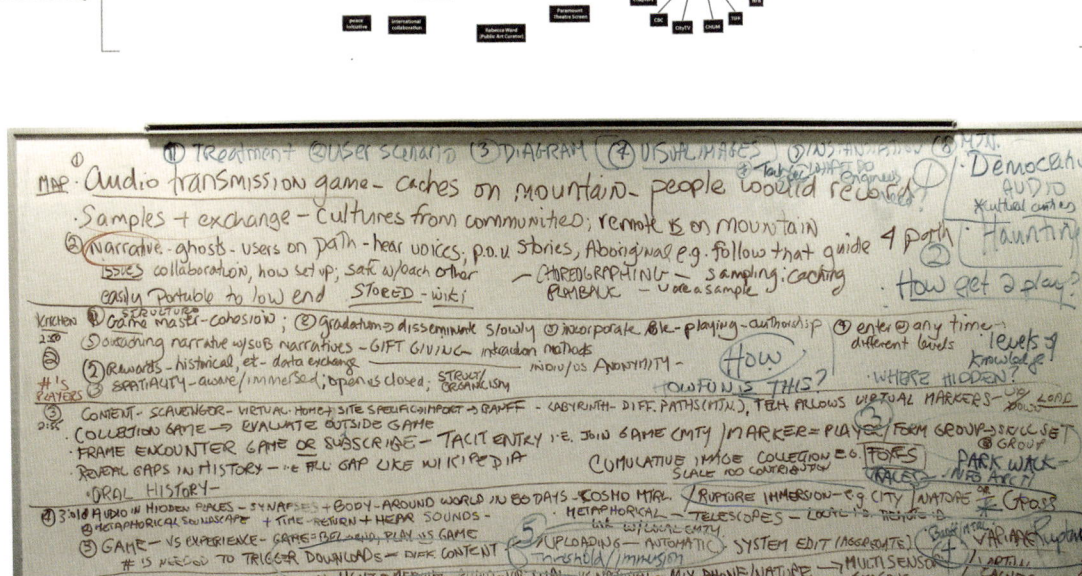

4 **Nigel Craig** testing audio application during a charette

5 **Paula Gardner** leading field testing

6 **Mark Poon** demonstrating audio application during a charette

Alter Audio and PORTAGE worked on two sets of overlapping but distinct requirements: the need for enabling technologies and the need for meaningful end-user experiences. With Alter Audio the research team began by imagining a number of capabilities that we thought would be useful to sound artists or other content creators. These were broken out into ten projected iterations ranging from simple (using the keypad to turn different combinations of sounds on and off) to complex (using environmental data such as GPS coordinates or altitude in combination with user actions such as congregating or gesturing to affect existing sounds or sounds input by current or previous users).

The engineering staff worked closely with the designers and artists on the team to build and test each new technical capability. Once an iteration became available, the artists and designers worked with engineers to create several different kinds of user experiences that took advantage of the new functionality. In many cases unexpected restraints emerged. For example, having other users enter into proximity (i.e. within ten metres of each other) triggered their sounds to play on each other's devices, but multiple sounds did not necessarily play in sync, making conventional music composition difficult.

At this point in the process we began to employ the principle of 'inside-out experience design.' The artists and designers, including students, professional composers, and the researchers themselves, would work within or around the limitations of the system, creating new works which often capitalized on what first appeared as a deficiency. In one case a designer, Nigel Craig, created soundtracks consisting of a series of single, sung tones which only took on an evocative nature when multiple versions were randomly combined.[5] In another, the designer created a series of complex sound arrangements with substantial periods of silence between them, allowing them to mesh with very different results depending on how they lined up. [6]

At the same time, a new demand would be introduced into the iterative development cycle, so that sub-iterations emerged. New functionality could be introduced, but existing functionality could be refined and adjusted based on feedback from the system's actual use (provided by both the sound designers and the end-users who were asked to test each experience).

Conclusion

In the designs we are creating we work from the middle of a set of capabilities and consider how those might be used to develop a unique and original user experience. At the same time, we interrogate the capabilities themselves, imagining how they could be retooled to respond to the emerging needs of the content.

In our work with Alter Audio, and more specifically our new work with PORTAGE, we adopt an inside-out approach to developing images, objects, and systems. Thanks largely to the availability of the Mobile Experience Engine (MEE) we have been able to start out with a linear, iterative development model that includes incremental additions of functionality to an interactive audio experience, and then deviate from that.[7]

Once a platform iteration has been achieved, we seek to populate it with various kinds of content, many of which strain the system and cause us to backpedal and make modifications before moving on in a newly informed, sometimes tangential, direction.

For example, we created *Phone Noir*, a two-voice, location-driven narrative which relied on Bluetooth beacons for its initial indoor presentation. When we were asked to present it in an outdoor venue as part of the national Words Aloud! poetry festival we were able to use the MEE to convert it to a GPS-driven application and introduce an element of synchronization.

Hence our use of the term 'inside-out.' We look outward to content producers to employ new creative vocabularies made possible by our platforms, but look inwards and revisit the platforms, tweaking them and making adjustments based on these real-world experiences.

Acknowledgments

Shea and Gardner thank our research associates, faculty Dr. David McIntosh and Patricio Davila; engineers Tom Donaldson, David Gauthier, Jagmit Singh, Sukhmeet Singh, and Ken Leung; research coordinator Brenda Goldstein; research assistants Nevena Niagolova, Jennie Ziemianin, and Peter Todd; production interns Mark Poon, John Pavicic, and Nigel Craig; and composers David Ogborn and Andrew Standlick.

Notes

[1] 'Alter Audio,' www.mobilelab.ca/alteraudio; the MDCN was led by principal investigators Sara Diamond and Michael Longford and funded by Canadian Heritage. See Mobile Digital Commons Network, 'Mobile Digital Commons Network,' www.mobilenation.ca/mdcn/

[2] PORTAGE is funded by Canadian Heritage, New Media Research and Development Initiative.

[3] The pseudo-democratic nature of participatory design is explored in Finn Kensing, *Methods and Practices in Participatory Design* (Copenhagen: University of Copenhagen, 2002) and Anne Marie Kanstrup, *D is for Democracy* (Aalborg: Aalborg University, 2003).

[4] Extreme Programming suggests that multiple, quickly developed solutions in software design result in more responsiveness to evolving demands or needs.

[5] This work is described in G. Shea and P. Gardner, 'Alter Audio: Mobile and Locative Sound Experiences,' www.mobilelab.ca/alteraudio/docs/Alter_Audio--Mobile_and_Locative_Sound_Experiences.pdf.

[6] The designer referred to here is David Ogborn. See G. Shea and P. Gardner, 'Alter Audio: Documentation,' www.mobilelab.ca/alteraudio/document.html

[7] MEE is a system which allows people with less technical skill than professional programmers to program applications for mobile use. Development was led by Tom Donaldson, Banff New Media Institute.

Warmware
mnemonic art and design research

Judith Doyle
Ontario College of Art & Design

When renowned designer and Ontario College of Art & Design integrated media graduate Robin Len was injured in a bike accident, I discovered first-hand how few resources exist to support amnesic individuals, who can find themselves confined in Alzheimer patient wards. My research led to Toronto's Baycrest—a centre for neurological research and innovation where neuropsychologist Brian Richards heads Memory Link, a support program to improve self-sufficiency and independence for clients with acquired brain injury. Storing and retrieving recently experienced events and information and acting on intentions at a future target time are vulnerable to both injury and aging. An inability to reliably create new memories following brain injury is called 'anterograde amnesia' and it poses devastating challenges, often rendering memory impaired individuals incapable of staying on track of intentions and activities or coordinated with friends and family. Because Memory Link participants experience problems storing and retrieving new memories they can be 'marooned in the moment,' as Carolyn Abraham writes.[1]

At the Baycrest Rotman Research Institute, memory operations are revealed by new neuroscience imaging technologies including MRI and PET scanners. Different regions of the brain are active and engage across a network in specialized tasks: language comprehension; emotional memory; recall of events, facts, names, and faces; procedural skills acquired through practice such as playing a musical instrument or tool use. These competencies are executed in different parts of the brain that reinforce each other. In the event of acquired brain injury, many types of memory and intelligence are undamaged and can compensate. The brain's undamaged abilities, including the procedural memory system for learning a new skill with practice, can be enlisted to train amnesic individuals to successfully use reminders, such as handheld personal organizers.

For over a decade, Baycrest research has focused on memory and cognitive supports. Richards' innovations include a 'memory book.'[2] This diary and address book has a detachable buzzer alarm that, while effective, was often embarrassing in public, like an obnoxious cellphone ring tone. The next iteration was the Orienting Tool, specialized software for Palm Pilot developed by Richards and Mike Wu, a graduate student in computer science at the Knowledge Media Design Institute (KMDI) who works closely with Brian Richards' in the development of memory aids. As part of the Toronto

Academic Health Sciences Network, Baycrest is affiliated with the University of Toronto, supporting research, teaching and clinical care. The KMDI includes Ron Baecker, Mike Wu, and their colleagues. The Orienting Tool that Wu and Richards developed is still used by Memory Link clients. It is distinguished from most reminder software by an amnesia-specific user-participant design methodology and the intensive training techniques Richards et al. developed to enlist users' procedural memory systems to recall how to operate the PDAs.

As a high-tech business accessory, a Palm Pilot doesn't signal itself as a cognitive prosthetic in social situations. A contrast between dependency (reliance on prosthetics) versus agency (user of new gadget, business tool, social networking, or entertainment device) is a factor in amnesic individuals willingness to use memory aids in daily life.

Warmware: An OCAD-Baycrest Research Initiative

The OCAD-Baycrest Research Initiative is rooted in a classroom-based collaboration engaging students' art and design expertise in the development of digital memory aids. A research collaboration entering its third year, the artists and designers contribute expertise in concept development and representation to address the emotional dimensions of amnesia. In preliminary discussions, Brian Richards and I considered emotional experience following brain injury (especially autonomy needs) and the emotional dimensions of disorientation experienced by individuals with anterograde amnesia. Not only are important past and upcoming events forgotten, but the feelings surrounding these events are lost.

We devised a three-stage intensive research, development, and documentation project over six weeks with approximately fifteen interdisciplinary students of the Virtual Communities class, two OCAD team-teachers, scientists from Baycrest and KMDI, and two Memory Link user-participants with their care workers.

We began by comparing research approaches each year, exchanging lectures between institutions. In my 2007 lecture to the Baycrest psychology rounds, I articulated differences between art and design methodologies in order to enrich our interdisciplinary conversation, and to clarify why art and design epistemologies must be included at the onset of the research and iterative processes with scientists and engineers. I outlined methodologies and vocabularies deployed by artists and designers including techniques for assessing and representing embodied memory, and perceptual expertise that could be enlisted to enrich memory scaffolding.

In developing 'warmware' my user-participant research model has been informed by Lars Erik Holmquist's theory of the 'extreme user.'[3] (Holmquist is at the Swedish Institute of Computer Science and heads the Future Applications Lab, Viktoria Institute, Göteborg.) Brain-injured clients of Baycrest's Memory Link program experience extreme impairments storing and retrieving memories of both events and surrounding emotions. Following Holmquist's theory, the novel insights obtained from OCAD and Baycrest's collaborative research to offset anterograde amnesia (problems remembering the future) can be generalized for a much larger user base (people with age-related mild cognitive impairment (MCI)).

Holmquist's model parallels 'extreme character' interaction design, and has been deployed in Phoebe Senger's autobiographical design theory.[4] These novel descriptions of the user-subject resonate with what I view as OCAD's strengths in art studio practice, critique and design research and development strategies. In these models, the people we work with may not be the end-users of the proposed system, but rather those who prompt interesting and novel insights.

Research Methodologies: Classroom Intensives and Assignments

The warmware collaboration and student project development takes place over the final six weeks of the thirteen-week Virtual Communities semester. This compressed development period parallels in some ways that of the design charette as an iterative methodology. David McIntosh describes the charette process in his paper in this anthology: 'The charettes involved collective participatory design and rapid prototyping, where brainstorming, scripting, asset production, interaction engineering, and on-site testing were all accomplished in six-hour periods.' The usual delay between concept, testing, and reiteration was drastically reduced.[5] Using collaboration techniques, brainstorming, and location-based bodystorming, the students and researchers rapidly move from concept to outline to rough version and revision. For warmware project development, assignments are often built as websites with blogs, audio tools, podcasts, and basic interaction programming. We make use of online maps and calendars, and social networks including Blogger, Facebook, YouTube, and *Second Life* as environments to test-pilot concepts. What follows is a brief summary of research themes and assignments.[6]

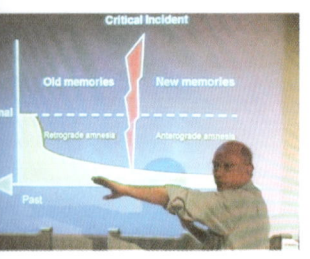

1 **Brian Richards**, psychologist and director of Baycrest's Memory Link program, lectures in OCAD's Virtual Communities class

Intensive six-week development periods include research lectures at OCAD from Brian Richards, Mike Wu, and Ron Baecker; a review of previous OCAD warmware research; and introductory discussions with Memory Link user-participants and their support personnel (*fig.1*). During this period, students are acquainted with their areas of experience and expertise as well as some of the common cognitive impairments of anterograde amnesia.[7] All assignments are posted on a class blog, as well as student photos, images from designs, audio and video files, comments, and other materials. The blog forms the core of a weekly context review. It is a social network and memory structure in itself, easily accessible online and in class to the students, researchers, and Memory Link user-participants and their support personnel, who also attend classes to test and give direct feedback on our findings and project outcomes.

  1. Create a Set of Multimedia 'Emoticons' (Reminders for Feelings)
     This assignment is designed to address the problems individuals with amnesia experience in remembering emotion. Not only are important events forgotten, but the feelings surrounding these events are lost.

Students select from a multimedia toolkit to create a family or group of emotional signifiers, keeping in mind these may be adapted for PDA's and handhelds. Examples have included sets of line drawings (Goldstein), video clips (Wright, Morris), photographs of facial expressions (Maybee), and graphics-stickers (Leung).

2. 'Just a Walk': Signposts Along a Memory Map
For this assignment, students create a series of 'signpost' drawings, photos, words, and audio or video clips leading from OCAD to a nearby real-world destination. Solutions have included an 'iSpy' rhyme with colour-coded signposts (Payne), software for a signpost updater (Morris), and a PDA-based sequence of zooms and spot-colour fades.

3. Adaptation into Documentation and Instruction Formats
During this period, students convert the previous assignments into a set of ingredients and sequence of steps, so that Memory Link clients and their support workers can create customized versions of, for example, sets of emotional signifiers. Instructions for doing so are posted on the blog.

4. Non-linear Timelines and Assistive Memory Devices
Here, students create accounts of an experience, event, or memory (and its emotional content) that occur over time, electing to work with words, image/texts, sound recordings, video clips, or images. The non-linear timelines can be shaped like a leaf, a sandwich, a pile of clothing—anything but a straight line.

In the subsequent assignment, students use the non-linear timelines as prototypes for assistive memory projects. The emotionally rich reminder systems might include text reminders, sounds or music clips, photos, drawings, video clips, interactive characters (avatars), or a mix of these. The assistive system may help to provide directions (an interactive map) or keep track of a list of things to do (like shopping lists) or may be structured as a calendar that reminds the user of appointments. As well as helping with planning and remembering the future, the device is an archive (a diary or photo album) that helps keep a record of feelings (emotional responses) as well as events.

For the assistive memory projects, students are asked, Who will use your system? Is it just for one personal user? Will family members and friends share this system, or a virtual community linked by technology that might include doctors and caregivers, or online friends? Projects are developed with in-class feedback from classmates, instructors, and our experts and user-participants from the Memory Link program.

Assistive projects have so far been designed for handheld devices (Palm Pilots, cellphones, and MP3 players), websites, alarm-clock type timers, and other devices. Some include wearable elements (Mistry). Most involve a combination of all these. Projects have included a handheld drawing tool (Temel, see *fig.2*), a mini-keyboard for tune notes (Fowler), remAUDIO reminder audiosharing (Singh), a memory 'ripple' map for organizing per-

2 **A warmware experiment** in memory, drawing, and process by Furkan Temel, a student in the OCAD course Virtual Communities

sonal photographs and contacts (Fung), a goth-style animé haunted house interface (Tsai), a navigation system using lengthening shadows, based on traditional African storytelling (Acheampong), a buffet of food-based digital metaphors including a chocolate box (Chen), refrigerator (Debi), and olive jar (Whetter), and a movie-style ranking system using hues and stars (Fowler).

Research Directions for the Future

Architectural memory support has a long tradition (in digital media, the Architecture Machine Group at MIT is one example). In the future, we will explore novel uses of gaming environments to enhance memory. For example, my own interactively navigable Foxscape project, created during a 2005 CanWest Global Fellowship at the Banff New Media Institute, uses video-game programming and 3-D modelling as a context for archival and family photographs. I am presently exploring immersive environments and 3-D installation of this gaming project to explore potentials of projected and built materials. (Doyle, see *fig.3*).

In the future, our OCAD-Baycrest research projects in Virtual Communities will include a memory architecture project using computer game software (possibly to be constructed on *Second Life*). Our research has shown that online user-driven media networks and virtual communities can provide helpful and emotionally rich mnemonic functions for brain-injured users. We will explore interactive planners with auto-archiving features (programmed passage from short- to long-term memory over time) using online calendar tools and social networks such as Facebook.

In spring 2007, we built an online interactive calendar including Robin Len's family, friends, and rehabilitation support workers that has been a useful orienting tool for him and the group. At the University of Art and Design, Helsinki, researchers in social media are exploring how digital media create new functionalities, driven by the special nature of digital devices. Social media can support memory and orientation, as in the case of this simple online social media network for Robin.

For amnesic individuals including Robin, pressing needs remain for housing, work and study options that maximize self-sufficiency. Collaborative networks for coordination between agencies can be innovated, but tools for autonomy within these networks are also required. Self-sufficiency and autonomy are addressed in our research through development of self-representation features for new technologies and social networks. We approach problems storing and retrieving new memory through creative timelining and dynamic archiving, as well as development of new digital memory architectures. Entering year three, the OCAD-Baycrest research initiative is an instance of art and design university researchers innovating in the memory link network by piloting new user-participant models for collaboration and invention.

facing page
3 **Judith Doyle:** My family home as memory architecture, built in the *Unreal Tournament 2* game engine with Maya 3-D software at the Banff New Media Institute, 2005

Acknowledgements

Artists and designers contribute to the OCAD-Baycrest research initiative now entering its third year with the development and iteration of memory aids, working in collaboration with neuroscientists, brain-injured users, and computer programmers. Baycrest psychologist Brian Richards, with programmers including Mike Wu at the University of Toronto's Knowledge Media Design Institute (KMDI), developed PDA software for memory and orientation aids for individuals experiencing amnesia. The students in the Ontario College of Art & Design (OCAD) Virtual Communities classes have collaborated with these scientists to bring art and design skills to this research.

Notes

[1] For an account of amnesia and Dr. Brian Richards's Memory Link project at Baycrest, see Carolyn Abraham, 'Marooned in the Moment,' www.innovationcanada.ca/18/en/articles/carolyn-a.html.

[2] For a full account of the development of the 'memory book' see Mike Wu, Ron Baecker, and Brian Richards, 'Participatory Design of an Orientation Aid for Amnesics,' (paper presentation, Conference on Human Factors in Computing Systems 2005, Portland, OR, April 2–7, 2005).

[3] Lars Erik Holmquist, 'User-Driven Innovation in the Future Applications Lab,' *Extended Abstracts of Conference on Human Factors in Computing Systems 2004.* (New York: ACM Press, 2004). 'Holmquist distinguishes his method from most participatory and ethnographic approaches. This distinction is pivotal in understanding potential generalization of warmware research... We are interested in cutting-edge technology, and in many cases, the initial idea for a project will be based on technical possibilities rather than any particular user need... Furthermore, unlike participatory design and most ethnographic approaches, we do not (necessarily) regard the groups we work with as the final users of a proposed system. Instead we see them as a springboard that will help us to push our ideas further. For this reason, we are interested in finding users that have very particular and perhaps peculiar requirements. We believe such specialized groups are more likely to put our technology in a new light, thus giving rise to interesting ideas... We can think of them as "extreme users," an analogue to the concept of "extreme characters," which are persona that are created to generate ideas in interaction design. As with extreme characters, the purpose is to inspire novel ideas that can be generalized for a larger audience. In several instances we have seen how the insights gained from working with specialized users has pushed the original technology and concepts much further than would otherwise have been the case.'

[4] Phoebe Sengers, 'Autobiographical Design,' (workshop presentation, Conference for Human Factors in Computing Systems 2006, Quebec City, PQ, April 22–27, 2006).

[5] '...the charettes provoked new expressions of imagination and improvisation through their intensity. The charettes involved collective participatory design and rapid prototyping, where brainstorming, scripting, asset production, interaction engineering, and on-site testing were all accomplished in six-hour periods. The usual delay between concept, testing, and reiteration was drastically reduced to

produce an even more intense dynamic across medium, mobility, site specificity, and virtuality where improvisational responses proliferated.' David MacIntosh, 'Being There: Uncanny Medium, Methodological Multiplicity, and Proliferative Embodied Creativity in *The Haunting*' in this volume. See also in this volume Martha Ladly, 'Research and Design for Mobile Platforms: A Walk in the Park' and Geoffrey Shea, 'Inside-out Experience Design.'

[6] The blogs can be read at http://ocad-virtualcommunities.blogspot.com for the 2005 project and http://virtualcommunities-ocad.blogspot.com for the 2006 project.

[7] For example, in the home, a lack of memory (anterograde amnesia) can lead to:
- misplacement of objects like glasses, papers, keys, telephone messages
- inability to complete day-to-day chores (being distracted and forgetting to return to a task that has been interrupted)
- going to a room but then forgetting why upon arrival
- not knowing what is coming up next
- not varying the choices of food eaten each day (which can be the same for many meals)
- inability to recall plot from a novel while reading it
- further instances, such as these suggested by Mike Wu, are listed on the Virtual Communities blogs.

Deep Places
mobile 2.0 and spatial experiences

Jan-Christoph Zoels
Experientia

This paper documents selected Web 2.0 and mobile trends and developments in social software and networking. It explores Mobile 2.0 user experience through proposed interaction behaviours, service models, and content areas within mobile spaces. Contextual interviews and ethnographic observations of teenagers and young adults in Turin, Italy, were made in the summer of 2006. The focus of this spatial experience exercise was to identify significant places and spaces, and to understand behaviour and current product usage.

Our observations from these interviews include

+ A constant search for the 'new'—trendy locations, 'something different.'

+ An importance placed on personalization, be it of MSN settings, mobile phones, or backpacks.

+ A difficulty integrating various devices, with technological devices mainly used for their primary function, e.g. connecting a mobile phone to a PC is perceived to be cumbersome, and digital cameras and MP3 players are not replaced even when their owner has more advanced mobile phones that have these functionalities.

+ A wish to easily access high-quality technologies while on the move.

Background research and contextual inquiry supported five broader observations.

1. Contextual Experiences: The Power of Mobile 2.0
   By understanding a person's context—time, place, people, events, use—Mobile 2.0 applications can provide more meaningful and relevant results or experiences (e.g. in searches or in games). Context also provides a focus on user-authored content, which is uploaded, tagged, and shared within Mobile 2.0 systems. Opportunities lie in facilitating user-authored content and in helping users to identify and locate content.

2. Awareness of the Other: Contextual Presence
   Mobile 2.0 applications are aware of both the presence of others and their context of use, and may help us to negotiate the 'other': Are they nearby? Do they belong to specific interest communities? Are they strangers or friends? Do they share interests? Are they available? Opportunities exist to inform users of the contextual presence of others and to manage access, particularly within always-on, always-present interest communities.

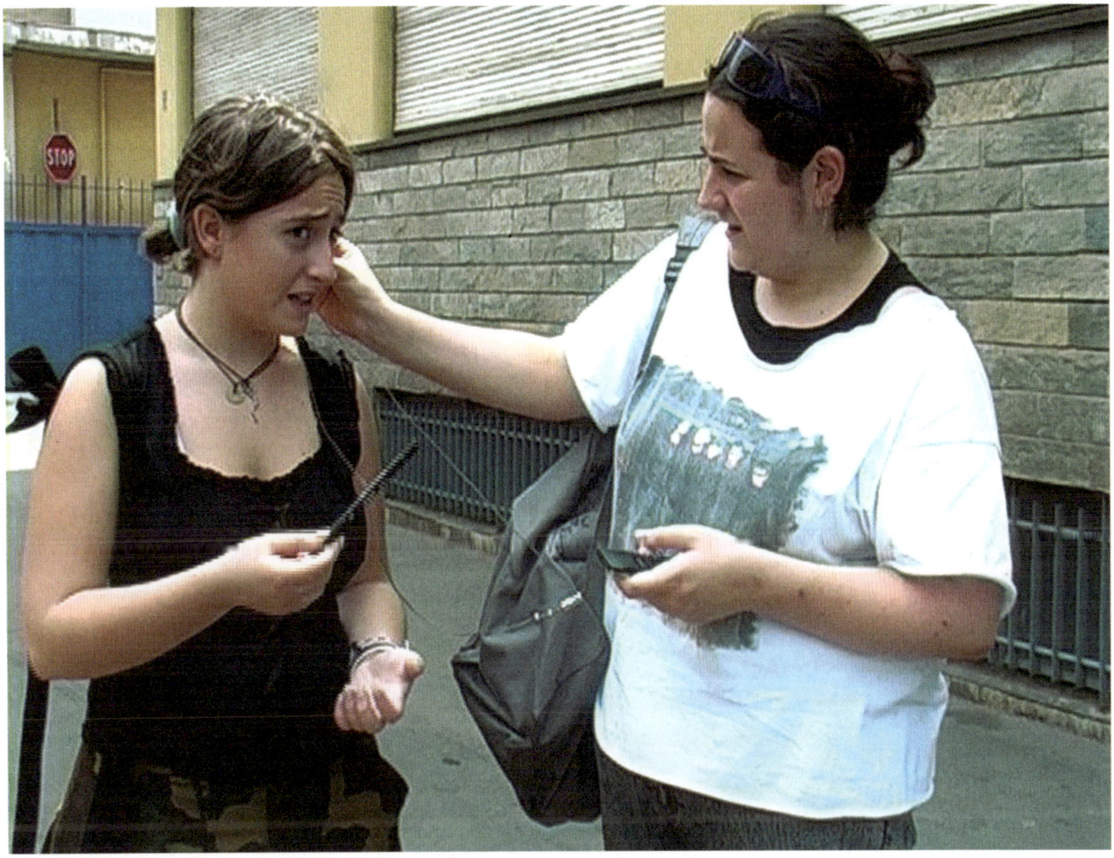
STOP

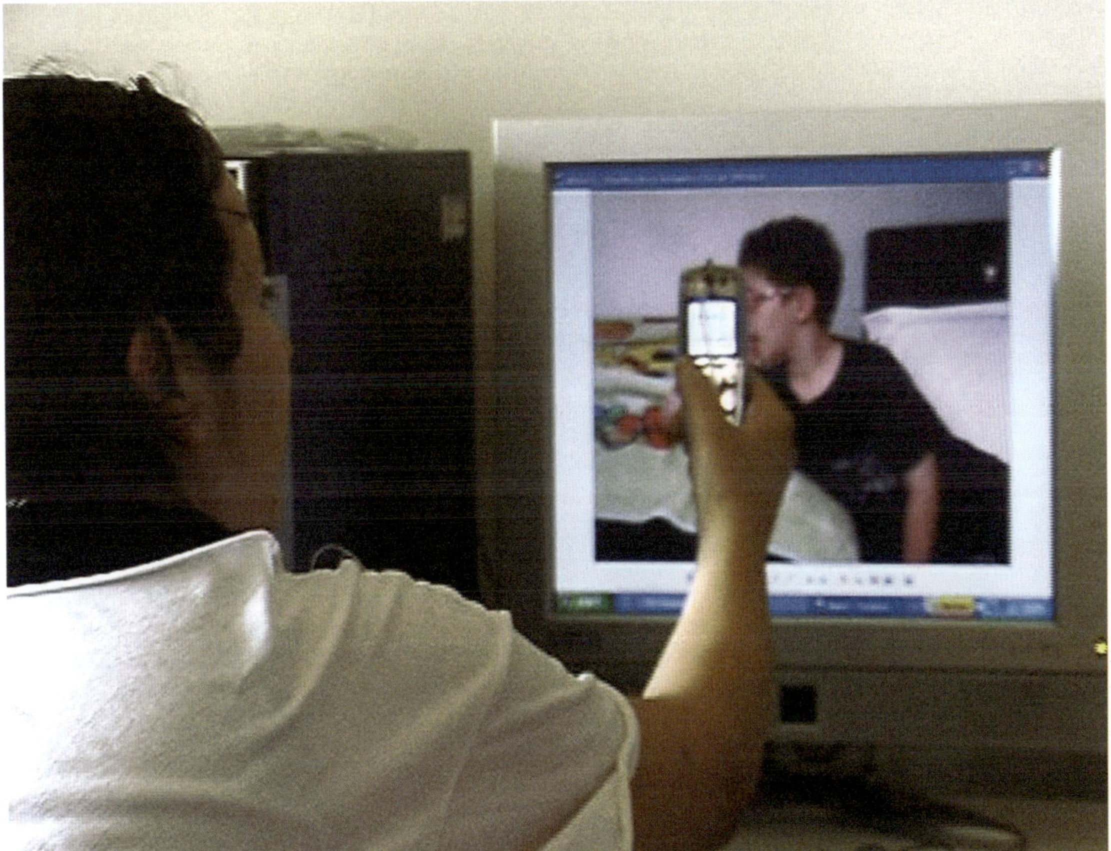

3 **A participant proudly** shows how she copies the most important SMS messages into her special SMS diary

3. Supporting the Real with the Virtual: Deep Spaces Concept
Combining the real world with the virtual enables Mobile 2.0 social networks and games to produce what we call 'deep space' or 'Space 2.0': an enhancement of the real world with information about interest communities or the textures of places, or with the increased engagement of a game. Opportunities could be combinations of these aspects, e.g. bringing context-specific interest communities into mobile game play, or tailoring virtual services at particular locations to meet specific needs of interest communities and individuals.

4. Complementing Web 2.0: Hi-res versus Timely Delivery
The higher information-resolution of web-based browsing will be coupled with the timely delivery of abbreviated information to a mobile device. Web 2.0 will not be eclipsed by Mobile 2.0 within a context of enriched spaces (Space 2.0); rather, both platforms will balance their strengths and weaknesses to provide a holistic and complementary location-based experience. Opportunities exist to extend this complementary location-based experience to such areas as social networking or gaming, and for delivering mixed web/mobile experiences in Space 2.0.

5. Exploring the Power of Many: Service Interactions Across Multiple Platforms and Locations
Mobile 2.0 applications are device independent, enabling access and participation anywhere, anytime. Such availability poses new challenges for users, who must now engage with service interactions across multiple platforms and locations. Opportunities lie in providing seamless user experiences and transitions between platforms of services, such as ensuring constant access to social networks and authoring tools for user-generated content.

Conclusion: What Does All This Mean for Locations?

Welcome to the concept of 'deep place.' Our surroundings are no longer just about what we can see with the naked eye. With the right tools we have the ability to discover great discounts around the corner, the little-known history of that statue just to our left, or maybe even to find a new love, who just so happens to be sitting on our right.

The use of public spaces observes a reciprocal relation between the properties of the space and the characteristics of the users that populate it. On one hand, the identity of a place is defined by the groups and activities that use it. On the other hand, individuals build their own identity based on where they go and how they behave in relation to the social norm of the place.

The use of mobile technology and ability to permanently be in contact has also changed the traditional concept of place and time. Appointments are now made in a fluid and less granular way. A square or street can be more interesting waiting places than a corner or a statue. The definition of place and time tends to be detailed over time as SMSs and phone calls are sent back and forth.

Another aspect is related to the negotiation of selecting a meeting point and the ability to describe it. Places with a short description that are easy to identify are more likely to be chosen as meeting points even though the group will tend to move to other places afterwards.

The use of public space and the decision of where to go are often associated to dominant personalities within the groups who determine and initiate new tendencies, ones that gradually influence the entire group. This so-called flocking phenomenon can be observed around music bands, sport activities like skating, etc.

Very often the characteristics of a place impact the way people behave and, consequently, the way they use mobile technology. For example, restaurants were described by some of the people we observed as places where we dedicate ourselves to another person. Therefore the interpersonal interruption by a phone was seen as disrespectful behaviour to another person in that context.

The aforementioned 'flocking' phenomenon can also be seen in the way people behave when consuming Internet, and in particular Web 2.0, tools. The decision-making process within groups is mimicked, for example, in the patterns of contribution online.

The resulting concepts of this study (currently under non-disclosure agreement with a major consumer electronics company) were grouped into the following categories: place making, tacit awareness, flocking, social gaming, community building, media sharing, and user interface innovations.

Acknowledgements

These insights are the result of an interactive team work by Experientia: Mark Vanderbeeken, Michele Visciola, Alexander Wiethoff, Ana-Camila Amorin, Dave Chiu, Hector Quilhet, Victor Szilagyi, and project lead Jan-Christoph Zoels.

Roots Not Wires

or, why mobile nations are local

Drew Hemment

Imagination@Lancaster, FutureEverything

The issues facing curators and creative producers working with mobile media and in the intermediate zones between wireless cells deserve critical reflection.

One of the most compelling responses to the emergence of mobile media has been the field of locative media. This is here understood to refer not just to digital media that has been spatially tagged, 'location based services,' or the interest in collaborative digital mapping. Rather, 'locative media' is understood to refer more broadly to a response to the migration of computing beyond the desktop and office into the world around us. It may be distinguished from related fields such as pervasive media, ubicomp, and so on by the clarity with which it captures a particular truth of mobile media. At the moment the cable was torn from the socket, we began to look to the place where we stand. With mobile telephony you are free to roam, hence location can no longer be taken for granted. Instead, location has emerged as a technical and artistic focus. Combined with this, intentionally or as a side effect of protocol, we are all being mapped all of the time. Locative media encapsulates both the hope and the fear that this generates, without closing off any of the ambiguity so implied.

The view that if net art is the art of the Internet, then locative art is the art of mobile and wireless systems, was developed through participation in a number of key projects and events.

In the course of these activities, a taxonomy of locative arts was developed, one which was first proposed in a volume of the journal *Leonardo*.[1] This sought to draw together related interests in location and proximity, mapping and mobility, and so on, and proposed three main categories of locative art: mapping, ambulation, and geoannotation, each of which is further subdivided into documentary, expressive, and social.

Note

[1] Drew Hemment, 'Locative Arts,' *Leonardo* 39 no. 4 (August 2006).

1 **Loca stickers** leave a physical trace of digital identities

2 **Loca:** A person walking through the city centre hears a beep on their phone and glances at the screen. Instead of an SMS alert they see a message reading 'We are currently experiencing difficulties moni-toring your position: please wave your network device in the air.'

facing page

3 **RICHAIR2030** wireless 'lunch box' in Mobile Connections exhibition, Manchester, UK

4 **Loca nodes** use readily accessible consumer components

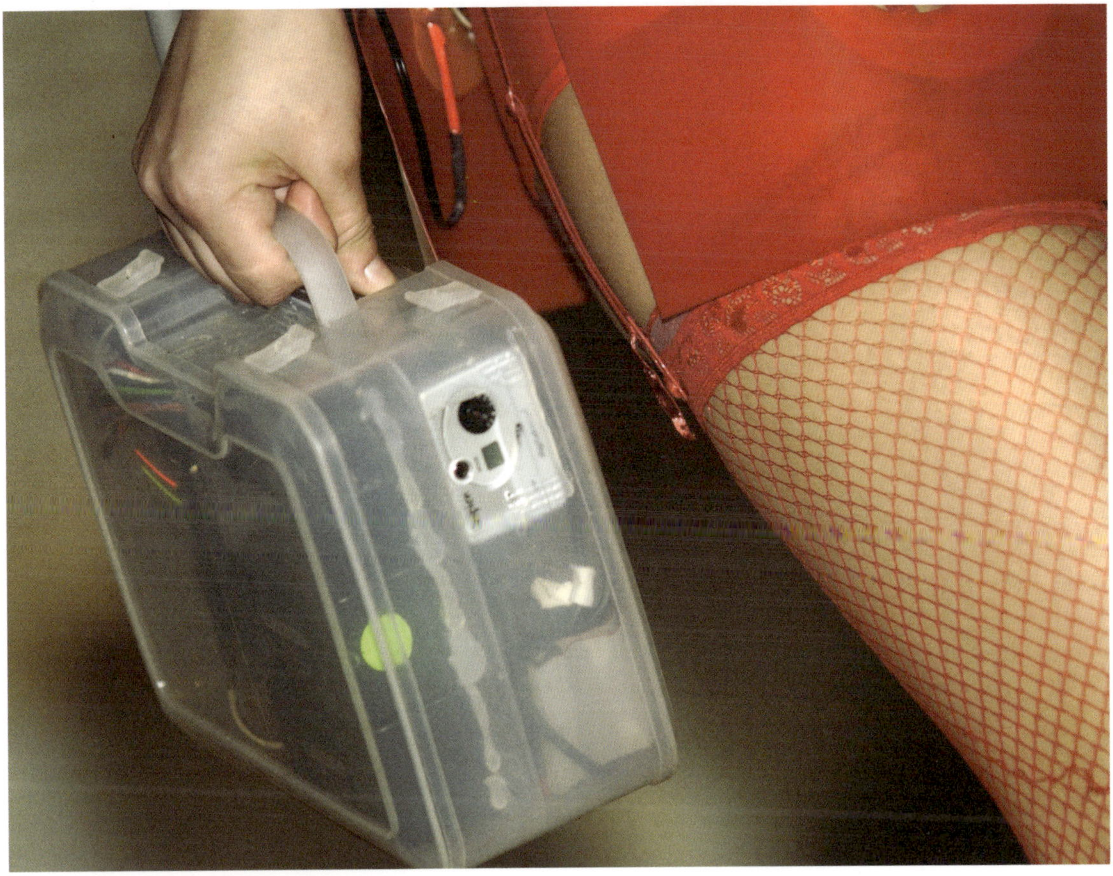

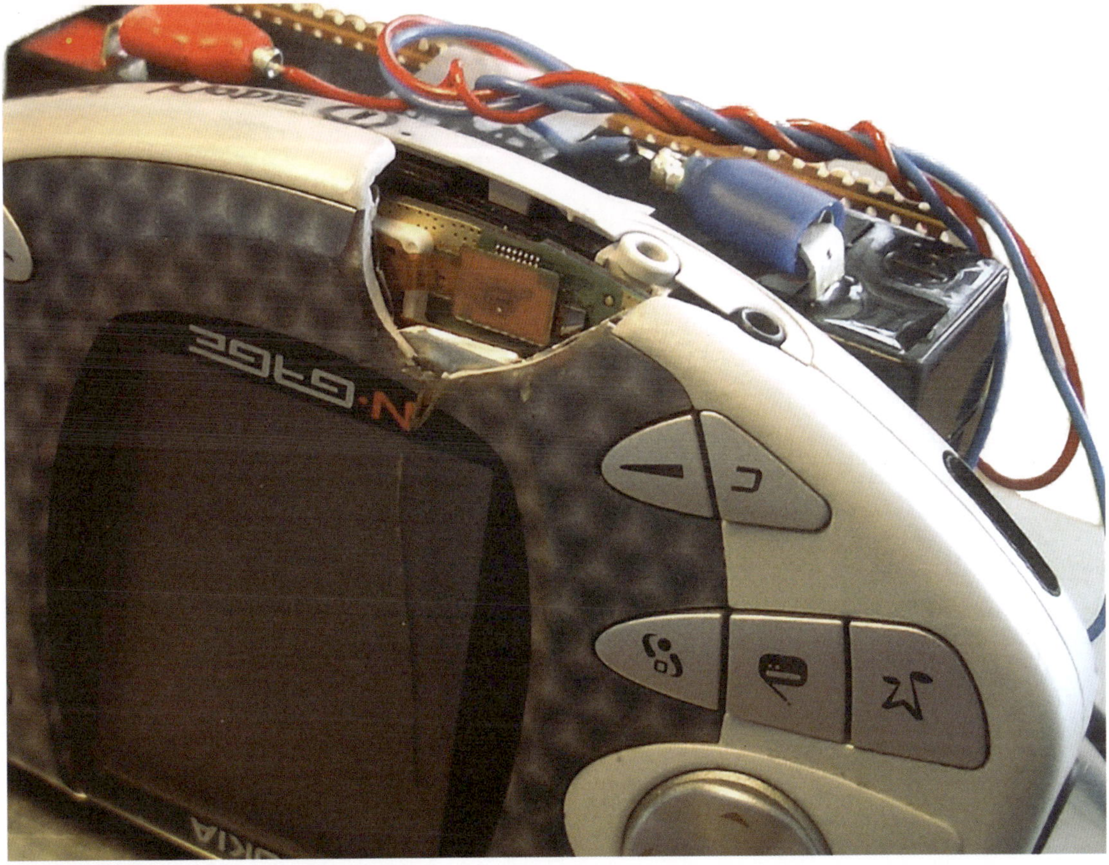

Engineering Meets Humanities and Social Science

Technological design needs to be conducted in close proximity to content and application design, requiring that artists, designers, and engineers develop effective forms of collaboration. Mobile Nation was interested in the following methods and research questions for engineering experience design. How do open-source formats work effectively with proprietary platforms and software designs? What are set procedures for mobile content design and engineering?

How can interoperability be designed and engineered within this evolving space? And where are the points of crossover with existing social science and other media research and design methods? Mobile Nation focused on the acceleration of these design and engineering methodologies, and compared the approaches of commercial research environments from Yahoo! Research, Nokia, and other providers of integrated mobile solutions, to the participatory methods used by Mobile Digital Commons Network researchers and in scholarly contexts, such as those at Carleton University, Concordia University, the Korea Advanced Institute of Science and Technology, Simon Fraser University, and the Ontario College of Art & Design.

The Mobilization of Art Practice
body metaphors and the desktop world view

Steve Daniels
Ryerson University

In 1997, Simon Penny published his influential paper 'The Virtualization of Art Practice: Body Knowledge and the Engineering Worldview.'[1] In that paper, he describes the collision of art practice with the underlying philosophies of digital technologies. He contrasts the holistic view-of-self and body-centred knowledge historically associated with artmaking with the Cartesian dualism embedded in digital computing by the engineering world view. Penny examines and questions the implications of this conflict in relation to the role of the body within this engineered digital space.

During the last decade, artists have actively subverted the relationships identified by Penny by co-opting off-the-shelf technologies, adopting DIY (do-it-yourself) strategies to interface design, and introducing networked, distributed, and ubiquitous practices. Reflecting on these changes provides insight into how the engineering world view maintains Cartesian dualism and how disrupting its standard leads to mobilization.

Bodies in Motion

Metaphors for the dislocated body of the dualist world view have long been aligned with dominant western scientific paradigms. In the time of Descartes, the rational myth that described universe and body was that of a machine.[2] Descartes' famous dictum, 'I think, therefore I am,' which led to the conceptual split between mind and body, both supported and propelled this belief. In the mid-nineteenth century atomized, reductionist thinking began to erode these long-held mechanistic beliefs. Developments in geology, biology, physics, and psychology redefined the universe and the body. However, it would take the publication of the structure of DNA by Watson and Crick, more than three hundred years after Descartes, for the mechanistic view to finally collapse. With the discovery of the structure of DNA, the shift in thinking that began with Mendel and Darwin became nearly complete. The dualist's body was transformed from machine to code. Significantly, the arrival of this new body metaphor is coincident with a braided journey of accelerating thought that entwines molecular biology, information theory, and digital computing, leaving the mechanical body behind.[3]

At least three responses develop from the convergence of this new plastic body and the dynamic space of computing. The material body in physical space, when seen as a manifestation of code, is subject to genetic modification and is eventually patented.[4] The material body is replaced

1 **DIY interfaces and works** from *Sources and Sinks*; a real-time networked event held in Toronto on March 31st, 2007, www.spinningtheweb.org

ENGINEERING MEETS HUMANITIES AND SOCIAL SCIENCE

in virtual space by avatars: code-based, online representations of the body that serve as identities for mental actions.[5] Finally, the lived-body now occupies a complex hybrid space between these extremes. This body contends with physical, virtual, and hybrid spaces by moving within and across these body metaphors in a context-dependent way. This body is machine, code, and mind. It is this flexibility and the contextual nature of response that DIY artists working with ubiquitous computing technologies are currently exploring. Interestingly, these new modes of hybrid computing are returning the lived body to geographic space.[6]

Mobilizing Models

Mobility = body + space + time

Communication = space + time

Mobility = body + communication

If the engineering world view affirms Cartesian dualism as Penny argued, then it must be recognized that the desktop model with its GUI is the animateur of this affirmation. This model defines a very limited body and a constrained space. The desktop forces a kind of *stasis* through layered simulation of space and by homogenizing 'a great variety of various bodily activities into one.'[7] The person that engages this computing space is a Cartesian machine that conforms to the desktop's demands.

New media artists have expanded computing practice over the past decade by moving beyond virtualization into hybridization.[8] In so doing, they have disrupted and displaced the desktop model. DIY interface design achieved through physical computing practice defines a richer relationship to the body and often hybridizes and activates space.[9] Combined with the relatively recent additions of networking and distributed systems, current practice pushes further away from the desktop model by introducing time. This is particularly true in systems where communications strategies (real time) are employed to disrupt and reconfigure space (i.e. telepresence, discussed below). It is telling that Penny does not include a discussion of communication or networking in his original paper. At the time of his writing these ideas had not yet been widely foregrounded in artistic practice.

Mobilizing People: Hybrid Spaces and Telepresence

Telepresent works attempt to juxtapose distant physical spaces using real-time communication networks and material interfaces (often robotic) activated or controlled by users.[10]

Since the spring of 2005, I have organized four telematic events involving Ryerson University's New Media area, InterAccess Electronic Media Arts Centre, SUNY Buffalo's Department of Media Study, and/or the Ontario College of Art & Design's Integrated Media area. These events have been tied to a variety of undergraduate and graduate classes in physical computing and/or telepresence at each of the universities listed. Each event has involved twenty-five to sixty emerging artists/students presenting unique networked interfaces and experiences situated in physically disparate locations.

Following Kac, I initiated these interconnecting projects thinking primarily about space.[11] I believed that collapsing space with real-time communication would be sufficient to create telepresent experiences. While this proved true in some technical sense, it became clear that space does not simply collapse. Generating contiguous space through real-time networks merges creative communities, juxtaposes users' bodies, and transforms space into a distinct hybrid reality. Telepresent audiences are not in one unified space, they are simultaneously in one and multiple spaces. As such the works depend upon and create participant movement as a way of understanding the consequent spaces. This movement is facilitated by body-driven interactions rather than traditional inputs and relies upon real-time, peer-based communication networks with strong feedback. Freed from desktop restrictions, participants in these spaces are social, mobile, and whole.

Telepresent systems provide unprecedented opportunities to reconfigure relationships between space, body, and communication and therefore provide a platform for actively questioning the desktop world view and its tightly bound experience. These distinct approaches to computing suggest simultaneously sliding scales of user experience along axes of body, space, and communication.[12] If these axes are arranged to describe a volume, we can place at its centre the desktop model described by Penny with its limited relationship to body and constrained space. We can move away from this centre, and therefore the desktop world view, by expanding along any or all of these dimensions. The telepresent events I have been collaboratively producing move away from this centre through their use of real-time communication, their unique interfaces, and their hybrid relationship to space.

These experiences have clarified for me the interconnections between space, body, and communication (culture/community) as they are entwined in our computing models. How we weave these elements can reinforce the engineering world view with its desktop GUI or allow us to produce experiences that stand in opposition to it. By taking a holistic view-of-self as a point of departure, it is possible to produce, within participants, a sense of social agency that is often lacking in the dualist model.

Acknowledgements

This work would not be possible without the continued support of the School of Image Arts at Ryerson University and the collaborative efforts of InterAccess Electronic Media Arts Centre, in particular its director, Dana Samuel. Each event depended on the tremendous efforts of countless students, support staff, and partners at other institutions. In particular I would like to thank project partners Marc Bohlen, SUNY Buffalo Department of Media Study; Doug Back, OCAD;

Johanna Householder, OCAD; Mike Stevenson, OCAD; David Green, Ryerson University New Media area; and Kathleen Pirrie-Adams, Ryerson University New Media area. This manuscript benefited greatly from comments by Caroline Langill and Kathleen Pirrie-Adams (some of which I ignored at my own peril).

Notes

[1] S. Penny, 'The Virtualization of Art Practice: Body Knowledge and the Engineering World View,' *Art Journal*, 56 no. 3 (1997).

[2] Daniel B. Botkin, *Discordant Harmonies: A New Ecology for the Twenty-first Century* (New York: Oxford, 1990).

[3] N. Katherine Hayles, 'The Condition of Virtuality,' in *The Digital Dialectic: New Essays on New Media*, ed. Peter Lunenfeld (Cambridge: MIT Press, 1999).

[4] S. Britton and D. Collins, eds. *The Eighth Day: The Transgenic Art of Eduardo Kac* (Tempe: Institute for Studies in the Arts, 2003); For a non-technical review of genetic modification and a history of gene patenting, see J. Rifkin, *The Biotech Century* (New York: Putnam, 1998).

[5] See for example http://secondlife.com.

[6] The work of Jane Jacobs makes an interesting intersection here. See also T. Dillon, 'Wired ways over calm cities,' www.futurelab.org.uk/viewpoint/art33.htm.

[7] Johnathan Crary, *Techniques of the Observer: On Vision and Modernity in the 19th Century* (Cambridge: MIT Press, 1992). In *Techniques of the Observer*, Crary describes a range of spaces generated by early optical devices. Today's desktop GUI with its flattened and layered space echoes the planar configuration of early stereoscopic images. The device for viewing these images suggests the head-mounted displays of early virtual reality work. The latter has been criticized for its immobilization of the user. For a history of the development of the GUI and its relationship to users see Bill Moggridge, ed. *Designing Interactions* (Cambridge: MIT Press, 2007), 521–537; Penny, 'The Virtualization of Art Practice,' 166.

[8] I am using *hybrid* in a way very similar to Hansen's use of *mixed reality*. The two terms point in different directions historically, but both recognize the uniqueness of these new spaces. See Mark Hansen, *Bodies in Code* (New York: Routledge, 2006).

[9] Dan O'Sullivan and Tom Igoe, *Physical Computing: Sensing and Controlling the Physical World with Computers* (Boston: Thomson Course Technology, 2004).

[10] E. Kac, 'Telepresence Art' in *Telepresence and Bio Art* (Ann Arbor: University of Michigan Press, 2005). See also K. Goldberg, ed. *The Robot in the Garden: Telerobotics and Telepistemology in the Age of the Internet* (Cambridge: MIT Press, 2001).

[11] Kac, 'Telepresence Art'.

[12] I am developing a visualization of the relationships between body, space, and communication . It can be found at www.spinningtheweb.org. Within this model the communication axis combines (confounds?) changes in the speed of delivery (movement towards real time) of information as well as the configuration of distribution systems (networks, not point-to-point broadcasts) that support that speed.

Hauntings across the Atlantic
The Marconi Trilogy

Leslie Sharpe
Indiana University Bloomington, Hope School of Fine Arts

I am haunted—haunted by tales of electronically charged change. Haunted by promises of formlessness, disguise, hidden identity. I can't shake the longing to 'beam up,' to pass through time and form, to function like a charged-up device without a shell. I am haunted by devices. Like a problem child carrying a blanket, I drag the device everywhere, stuffed in my pocket, clutched in my hand, inserted in my ear, [wondering about other options…]. My devices are haunted. These things that I carry project voices and words. I listen for their call. [They don't seem human.] What is this thing that wants to get out? My devices haunt houses and highways and horizons, endowing place with a new sense of space. They are here and there, and somehow nowhere. They create a new here where I am there. This haunting makes a sound. It crackles in the air. Silent movement, buzzing paths, humming networks, crackling connections, snapping disconnections.

For several years I have been working on projects that employ narrative and scientific metaphors of ghosts and haunted space as well as histories of wireless to address questions around embodiment and presence in our wireless age, and to explore new possibilities regarding authorship and distribution of the digital art object. The projects typically employ some kind of mobile device such as wireless PDA, cellphone, and iPod.

The mobile device, itself a technological object that can store and 'transmit' encoded and invisible data, suggests a temporal 'haunted' space and I treat it as such in my work. The *Marconi Trilogy* project also considers that the space in which electronic transmissions take place is full of mutable 'things' and asks what the signals or messages sent during transmission might encounter in that space and what we might consider embodiment or subjectivity in a wireless space. *Passing SG7777* and *Sending SG4L* are loose multimedia narratives concerning a ghost that is actually a signal—lost by Marconi in one of his first wireless transatlantic transfers—and how this signal collides in Hertzian and real space over the Atlantic with other ghostly forms (other electronic data, remnants of the old Marconi stations, flying kites at the station of Signal Hill, fairies, and ghosts of frozen bodies among the icebergs off Newfoundland). The narrative is also influenced by traditional folklore of the Atlantic region, which has a strong narrative tradition around those lost at sea that takes form in ghost and fairy tales.

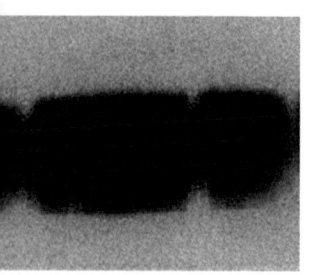

1 **The Marconi Trilogy**
2005–2007, detail from
*Passing SG7777:
Marconi Towers, Wellfleet*

facing page
2 **The Marconi Trilogy**
2005–2007, detail from
sg4l podcast website
for the exhibition Surge
curated by Rhizome and
Free104Point9, New York

3 **Ghosts for Cellphones**
2004–present, night
vision miniature images

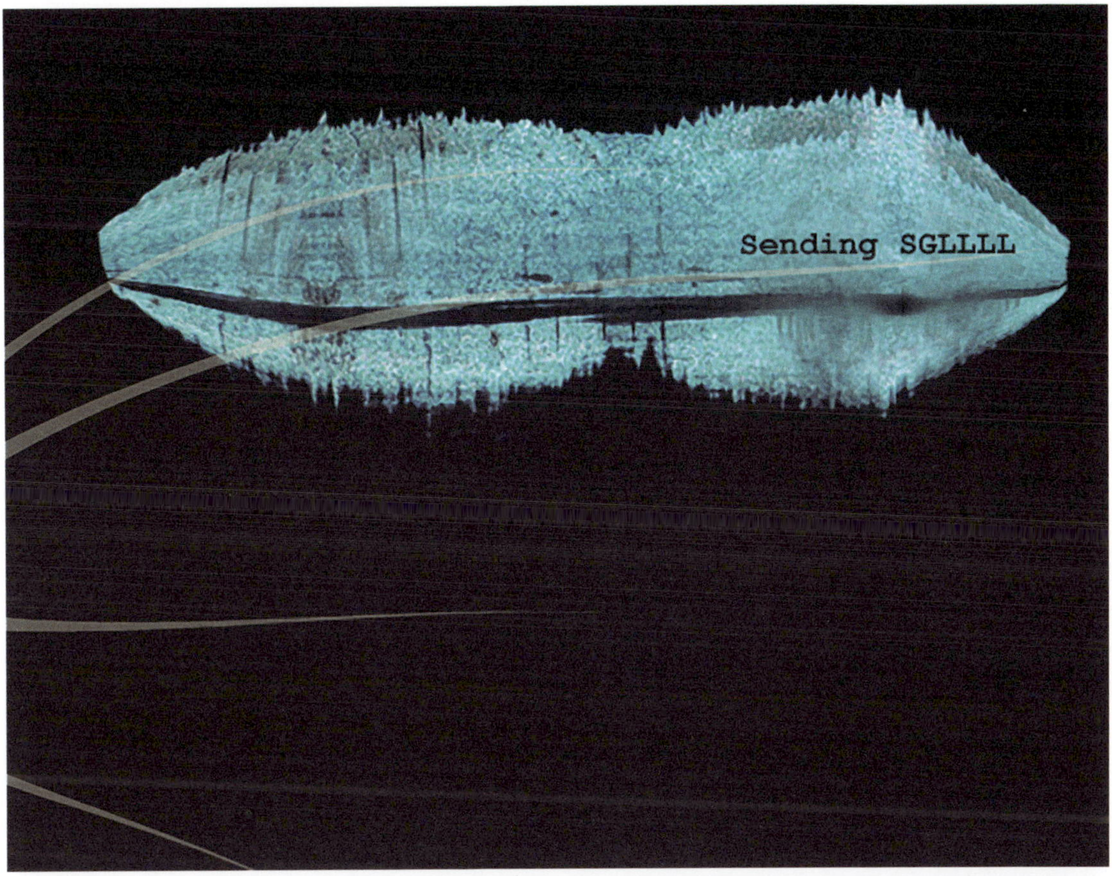

Sending SGLLLL

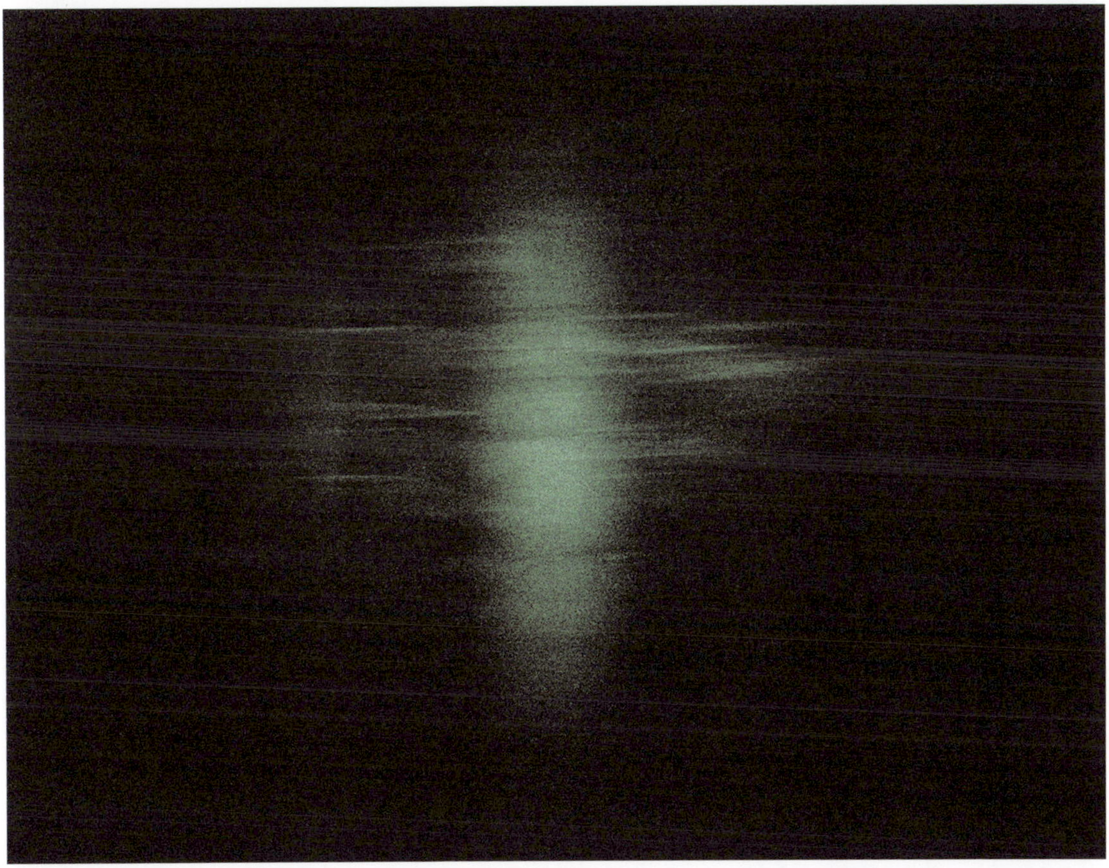

The title of *Passing SG7777* refers to the term 'sensor ghosts'—things that show up on radar screens but aren't really there—and to Guglielmo Marconi's 1900 application for patent 7777, a patent designed to reduce interference between stations by operating at different wavelengths. *Sending SG4L* upturns the lucky sevens, looking specifically for interference. 'Passing' and 'sending' refer to the distributive aspect of the works, as elements are transferred via Bluetooth to other Bluetooth devices, or broadcast as podcasts which can be then further disseminated into the public realm, i.e. as an image file sent from one's cellphone or a shared audio file. This dissemination invites unknown alteration by creating situations (in installation or performance) that encourage short-range transfer of files from and to my devices by nearby cellphones or PDAs. These situations can 'mess up' the original form of the work as others start to alter it by use of the devices, leaving the 'haunting' remnants of the audience as producer. This invitation to the audience to participate is a long-standing aspect of my work, and a means of seeking out new possibilities for the roles and groupings of audience and author, new forms of distribution and transmission, and new allowances for the 'art objects' we produce to be reformed in a dialogic fashion as they are disseminated and appropriated.

4 **The Marconi Trilogy**
2005–2007, details from
Passing SG7777:
Electronic presence at
Wellfleet

5 **The Marconi Trilogy**
2005–2007, details from
Passing SG7777:
Marconi Towers, Louisbourg,
Nova Scotia

facing page
6 **The Marconi Trilogy**
2005–2007, detail from
Passing SG7777:
Table Head, Glace Bay,
Nova Scotia

7 **Passing SG7777**
2005, detail from séance
performance and installation
using Bluetooth-enabled
PDAs, created for Blur of the
Otherworldly, Center for Art
and Visual Culture, Baltimore

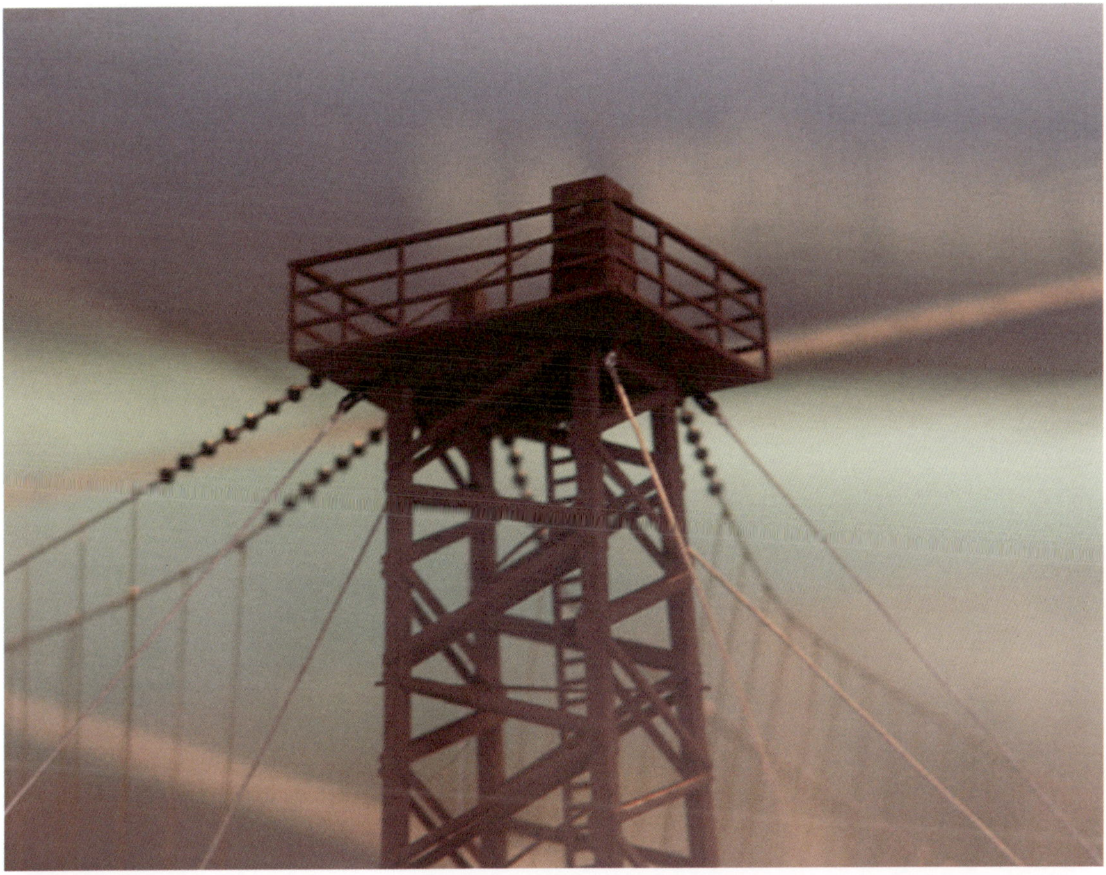

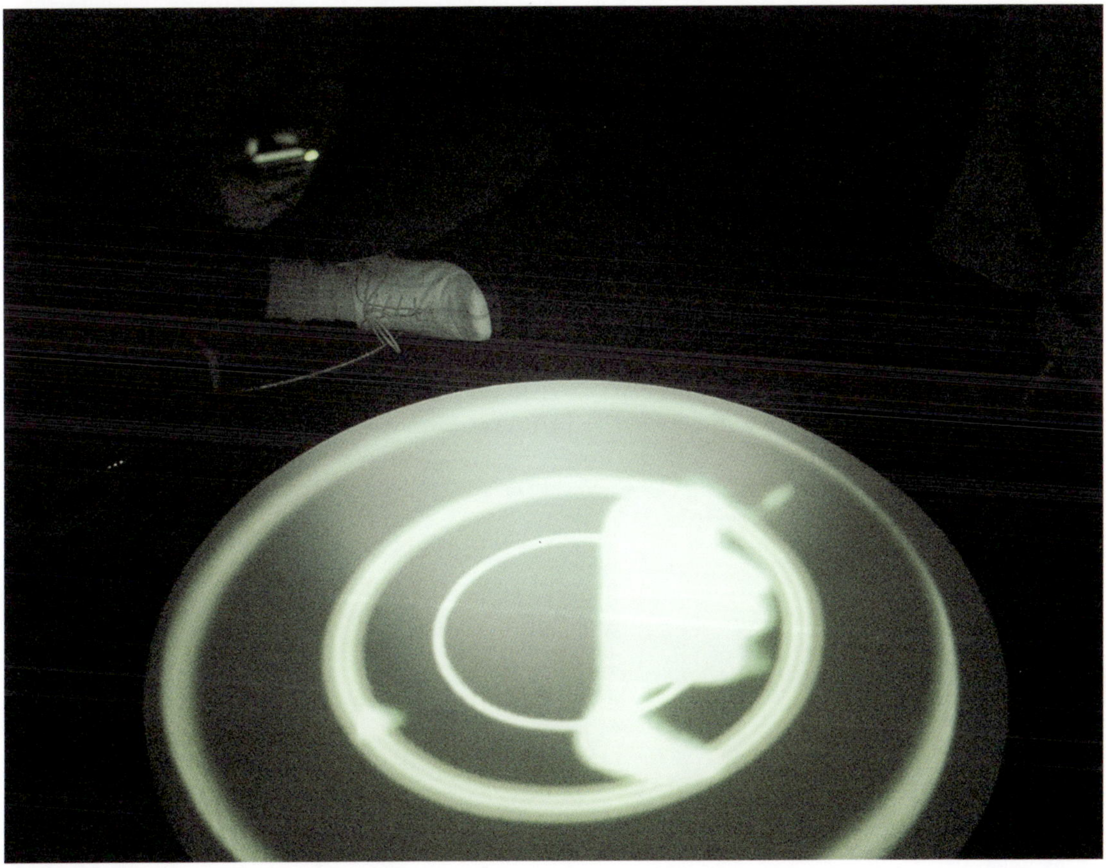

PageCraft

a tangible interactive storytelling platform to meet the needs of kids on the go

Jim Budd
Carleton University

Krystina Madej
Simon Fraser University

Jenna Stephens-Wells
Carleton University

Janice de Jong
Carleton University

Ehren Katzur
Carleton University

Laura Mulligan
Carleton University

PageCraft is a prototype framework for a tangible interactive storytelling platform that supports narrative development for young children. In designing such a system it is important to recognize the role of context as children grow and develop—children learn on the go. Historically it has been convenient to pack up books and building blocks for a summer vacation or a day trip to a grandparent's house. However, most technically oriented products don't meet these criteria. *PageCraft* is a multidisciplinary initiative that builds from theory and technical feasibility by combining experience in interactive product design together with research in digital narrative to bring narrative to 'children on the go.'

Emerging Narrative Development

In their play activities, young children take an interest in manipulating objects and imbuing objects with personalities. Referencing stories they have heard or creating new stories, they use objects to build story worlds during play with friends, parents, or on their own. As they progress cognitively, they conquer the semiotic challenge of relating objects and visuals to text and develop greater fluency with understanding a variety of narrative genres and authoring narrative. The inherent interest in giving objects meaning provides an opportunity for structuring a story-building system that is progressive and supports a young child's development in narrative and text literacy.

Tangible Interaction

The *PageCraft* system moves beyond the mouse and keyboard and provides children with hands-on interaction by using 'real' building blocks and character figures as the primary interface with the system. The use of physical artifacts to 'build' the story circumvents potential dexterity problems by replacing the point-and-click paradigm of the computer mouse with recognizable objects that a child can grasp and manipulate in a familiar way.[1] The use of a graspable interface has also been shown to encourage collaboration and increase interest, engagement, and understanding.[2]

In the *PageCraft* system *(fig.1)*, building blocks and shapes similar to the type young children use in play activities are sensor-tagged to work with story programs that 'read' position and translate placement and movement into onscreen text and graphics. Children build story scenes, manipulate and change the position of characters and props, record text captions, and then save images 'chapter' by 'chapter' to later print out. *PageCraft* also incorporates the use of ambient sounds, lights, animation, and elements of surprise including visual and audio feedback.

The system provides children a range of narrative experiences through a tangible user interface that leverages the capabilities of digital resources to provide a richer multisensory narrative experience.

1 **PageCraft system** supports storytelling activities with tangible narrative objects and ambient background media

Modes of Operation

PageCraft consists of a series of printed stories, a laptop 'book' with interchangeable playmats, a collection of blocks and characters with embedded sensors, and a carrying case *(fig.2)*. *PageCraft* provides continuity between traditional print materials, set-building props, and digital media by leveraging the benefits of tangible interaction. Printed stories allow for traditional storytime interaction between parents and children, and provide children with print materials to explore stories on their own.

The same stories can be 'read' by the electronic tablet in the form of an interactive digital storybook. The character blocks *(fig.4)*, which correspond to the characters in the stories, are part of a set of physical building blocks/story props/narrative objects with embedded sensors that a child can use to recreate scenes from one of the existing stories or to build an original story set on the playmat through solo play or in collaboration with a friend. By introducing the electronic tablet, a parent can read a story to the child while following along with printed text, or encourage the child to simultaneously build the story set with the *PageCraft* blocks. For more complex interaction, children can freely play with blocks and characters, then record accompanying dialogue to complement the scene they have just constructed. The recordings can be played back or archived and replayed in the future to share with parents and friends.

2 **PageCraft** can be played anywhere—in the home, in school, or at a friend's. The system consists of a series of printed books, an electronic tablet, and play surface, plus an assortment of colourful blocks, figures, and playmats all easily transported in a soft, flexible carrying case.

Users and Context

PageCraft encourages parent/child involvement while providing parents with an opportunity to engage in collaborative efforts with their children (*fig.3*). Like a child's book or set of building blocks, *PageCraft* is designed to be portable. The goal is to build a product that will appropriately fit the parent/child lifestyle versus a technical toy that demands a special location or environment.[3] *PageCraft*'s carrying case broadens the potential context of use, as it can be transported to school, to a friend's house, or even to visit grandparents.

How the System Works

Building blocks and figures (*fig.4*) contain sensors which identify their type and placement on the electronic tablet. As the child constructs a scene by placing the blocks and figures on the playmat, the *PageCraft* tablet identifies the placement of elements added or removed and generates the scene image from a databank of pre-built images. Once a child has built a scene he is happy with, he can choose to record an audio track. The computer converts the voice to text and adds a caption to the story page. A simple menu pad provides access to a number of choices for both recording and playback: the story can then be read out loud, printed for use as a storybook, or printed as a colouring book.

Summary

PageCraft is an exploratory prototype designed around an integrated interaction model based on child development/narrative theory. The objective is to develop the technology to respond to the narrative interaction model. Based on this approach, experiments are being conducted with different technologies in order to identify an optimal implementation. This is an iterative process. Various elements of the system are prototyped independently to support user engagement in the ongoing design development and testing process. Preliminary operational versions of various elements of the system employ different types of sensors to help build a hands-on understanding of the pros and cons of different technologies for this particular application.

Future Work

PageCraft is work in progress. Parts of a preliminary prototype (*fig.5*) have been built and are undergoing development and testing. The second phase of development will address issues of extensibility and include plans for a tangible editing interface for more advanced users. There are also plans to investigate the opportunity to add physical effectors and other tangible components to extend and enhance the narrative experience.

3 **PageCraft** can be used collaboratively encouraging socialization with parents, adult caregivers, and peers. For the child playing alone creating narratives provides a sense of agency.

facing page

4 **Colour, shape, texture,** and visual characteristics of the narrative objects help prompt the storyline. The building blocks and figures contain sensor tags which identify their type and placement in the scene generated on the electronic tablet.

ENGINEERING MEETS HUMANITIES AND SOCIAL SCIENCE

Notes

¹ G.W. Fitzmaurice, H. Ishii, and W. Buxton, 'Bricks: Laying the Foundations for Graspable User Interfaces,' in *Proceedings of the 1995 Conference on Human Factors in Computing Systems* (New York: ACM Press, 1995), 442–449; H. Ishii and B. Ullmer, 'Tangible Bits: Towards Seamless Interfaces between People, Bits and Atoms,' *Proceedings of the 1997 Conference on Human Factors in Computing Systems* (New York: ACM Press, 1997), 234–241.

² D. Stanton, V. Bayon, H. Neale, A. Ghali, S. Benford, S. Cobb, R. Ingram, C. O'Malley, J. Wilson, and T. Pridmore, 'Classroom Collaboration in the Design of Tangible Interfaces for Storytelling,' in *Proceedings of the 1997 Conference on Human Factors in Computing Systems* (New York: ACM Press, 1997); J.A. Fails, A. Druin, M.L. Guha, G. Chipman, S. Simms, and W. Churaman, 'Child's Play: A Comparison of Desktop and Physical Interactive Environments,' *Proceedings of the 2005 Conference on Human Factors in Computing Systems* New York: ACM Press, 2005).

³ D. Stanton et al., 2001.

5 **Image** of the preliminary prototype shows the storybook, the building kit, the playmat, and a laptop computer (in place of a dedicated electronic tablet)

facing page
6 **PageCraft** is a portable story-telling system that moves beyond the mouse and keyboard and provides children with hands-on interaction by using 'real' building blocks and character figures as the primary interface with the system. The system provides children a range of narrative experiences through a tangible user interface that leverages the capabilities of digital resources to provide a richer multisensory narrative experience.

Interactive, Tangible, and Augmented Prototyping with MIDAS

Tek-Jin Nam

Korea Advanced Institute of Science and Technology

Introduction

In the traditional industrial age, the main role of designers was to create aesthetic forms of mass-manufactured products. The recently developed networked digital society has changed this traditional design paradigm. Designers now add humane value to new technologies and create application concepts to enhance the quality of human life. In the industrial age, design mainly dealt with hardware parts of static (non-interactive) manufactured products. However, in the networked information society, the objects of design have been extended to hardware-software integrated, dynamic, and interactive products or services. Emerging technologies such as ubiquitous or wearable computing have accelerated the emergence of various kinds of products, systems, or services that are digitally converged and interactive. This shift of the design paradigm attracts many designers' attention towards interactive digital design projects, tangible media, and physical computing.[1]

One of the most important design tools for developing innovative interactive digital products is rapid and effective prototyping.[2] It allows designers to iteratively explore, develop, and evaluate new concepts. Prototyping is also a means of communication between participants in the design process. Traditionally, sketches or foam mock-ups have been used as such design tools for hardware-centric product concept development. However, for interactive digital products, design tools should also support the exploration of additional aspects such as contents, user interface, and integration of hardware and software.

However, designers without engineering backgrounds have difficulty in creating tangible and working prototypes in the early phases of the design process. Engineering development tools, such as programming languages and microcontrollers, are now introduced to designers. Some new tools have been developed specially for designers and artists.[3] However, they still have limitations in bridging designers' concept development and technical implementation, and are considered difficult for intuitive and flexible prototyping.

[1] H. Ullmer, H. H. Ishii, 'Emerging Frameworks for Tangible User Interfaces,' *IBM Systems Journal*, 39, no. 3-4 (2000): 915–931; D. O'Sullivan and T. Igoe, *Physical Computing: Sensing and Controlling the Physical World with Computers* (Boston: Thomson Course Technology, 2004).

[2] T. Kelley, T. J. Littman, T. Peters, *The Art of Innovation: Lessons in Creativity from IDEO, America's Leading Design Firm* (New York: Random House, 2001).

[3] Processing, 'Processing (Beta),' 2007, www.proce55ing.org; Cycling 74, 'Max/MSP,' 2007, www.cycling74.com/products/maxmsp.

facing page
1 **System architecture** of the MIDAS prototyping environment

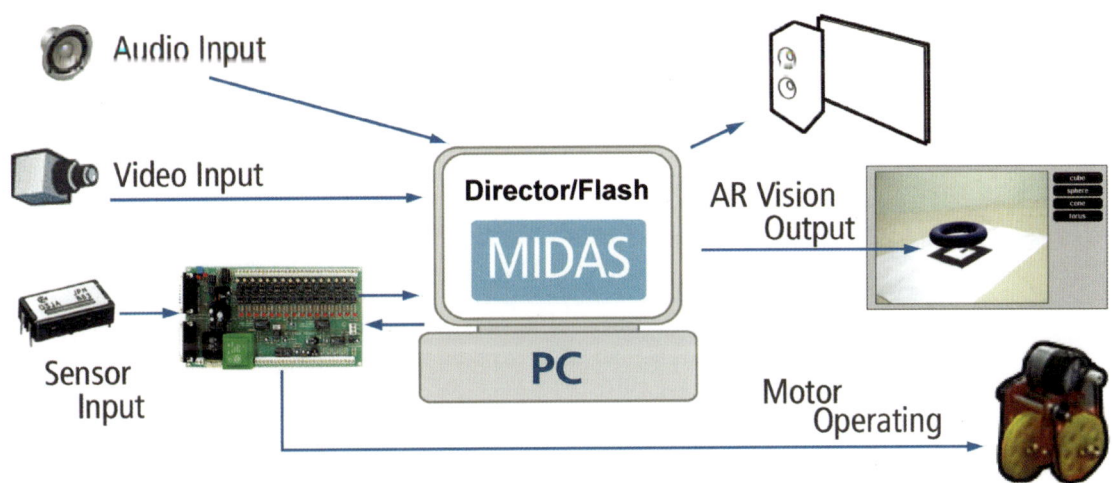

Audio Input

Video Input

Sensor
Input

Director/Flash

MIDAS

PC

AR Vision
Output

cube
sphere
cone
torus

Motor
Operating

MIDAS

We at the Department of Industrial Design, KAIST, have investigated more transparent prototyping platforms for designers. The Media Interaction Design Authoring System (MIDAS) is one of the prototyping tools that we developed and used for our design education and practice. It allows designers or artists with non-technical backgrounds to easily implement tangible working prototypes of conceptual ideas for interaction design projects, such as physical computing, tangible user interfaces, and interactive media artworks. Designers do not have to learn a new tool since MIDAS runs on several software applications, such as Macromedia Director, Macromedia Flash, and Microsoft PowerPoint, which are widely used among designers.

MIDAS simplifies the prototyping process with three parts of the interactivity cycle: listening (input), thinking (processing), and speaking (output).[4] It supports the simple connection of external input and output hardware to software applications running on a PC. It also supports computer vision and augmented reality within popularly used multimedia authoring tools such as Director or Flash (*fig. 1*).

MIDAS is a set of hardware parts and software plug-ins which includes several modules that can be used either together or individually. A typical MIDAS set consists of interface boards, input and output electronics parts such as sensors and motors, and plug-in software modules for Flash, Director, or PowerPoint.[5] It is mainly targeted to work within Director or Flash since those applications are the most popular interactive design tools for designers. Director or Flash users can easily control hardware inputs and outputs as if they control graphical objects in multimedia authoring tools. Lingo or ActionScript, embedded programming languages of Director and Flash, should be used together with MIDAS to add interactive features. Visual Basic for Applications is used as programming script in PowerPoint. MIDAS supports 3-D augmented reality by overlaying 3-D virtual objects on marker images in a live video captured by a camera in Director. MIDAS also supports simple vision-tracking in Flash. Those features make cheap PC cameras powerful sensing devices. It is useful to simulate interactive design concepts for envisioning user experience with new interactive systems.

Sample Projects

We used MIDAS for several design projects including mobile platform prototyping, tangible game implementation, and interactive media installation.

In the case of prototyping a wireless mobile device, RF modules were connected to the interface board, and electronic signals were transferred directly to the software simulation in Director or Flash. MIDAS was also used to demonstrate user experience in a design project about an information system solution for the blind. An interactive haptic cane and related mobile information service were suggested as the main concept of the project. Designers used the video tracking and 3-D augmented reality

[4] **C. Crawford**, *The Art of Interactive Design* (San Francisco: No Starch Press, 2003).

[5] **Velleman** 2007, www.velleman.be; **Trossen Robotics**, 'Phidgets,' 2007, www.trossenrobotics.com/hcihome.aspx.

facing page
2 ***Liquid Music Therapy*** prototype, an ambient multimedia product with a natural interface coupling the auditory element and the olfactory element of interaction, developed using the MIDAS vision tracking feature.

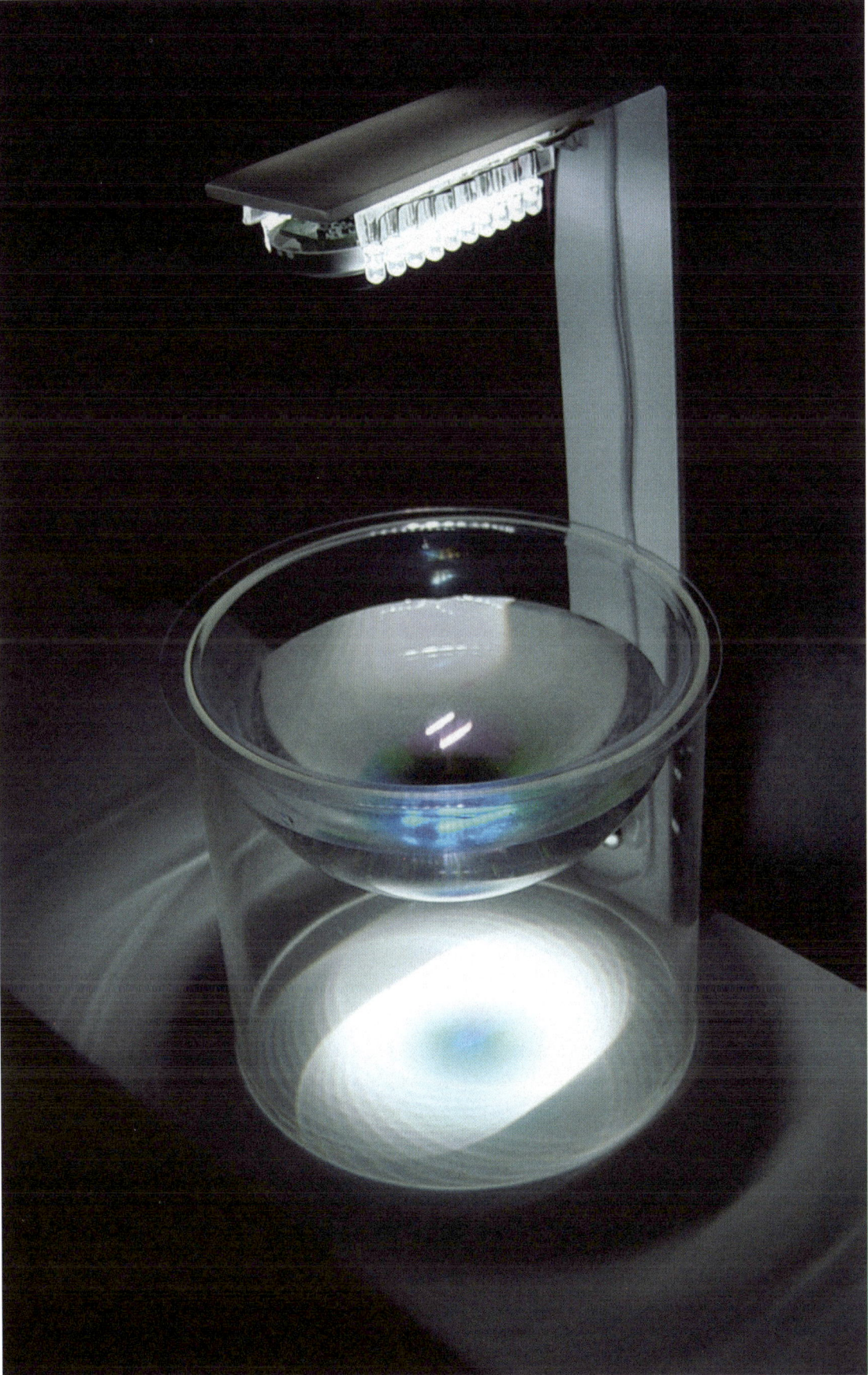

with a miniature street model to show the user experience. The haptic cane was prototyped as a full-scale working model. The combination of the miniature computer simulation and hardware device worked as an interactive tangible prototype that allowed designers to explore more design issues for the system development.

Liquid Music Therapy (*fig. 2*) is another example that was developed using MIDAS. It is an ambient multimedia product with a natural interface coupling the auditory element and the olfactory element of interaction. When people drop coloured, scented liquid into the device, it generates music and aroma. To implement this feature in Flash, MIDAS's video tracker was used to detect the size and the position of the coloured liquid in the water container. Tracking the colours can control music, which is harmonized with the colour and aroma. A PC camera was used for the implementation. In this procedure, a product designer with no engineering background could simply develop the sensing part of the product.

Students and designers who used MIDAS reported that it was very helpful as they could concentrate more on concept development. They could implement a wide range of interactive, tangible, and augmented prototypes without in-depth engineering skills. The time for prototyping was greatly reduced and the design process became iterative.

Conclusion

To reflect the new design paradigm in the digital society, effective prototyping tools for designers are a central concern for design education and practice. MIDAS attempts to provide one solution for interactive tangible and physical prototyping. MIDAS, and other relevant examples, are available at http://cidr.kaist.ac.kr/midas. In the Collaboration and Interaction Design Research Group at KAIST, we are also investigating other prototyping issues, including how to prototype software parts of interactive systems (*fig. 3*),[6] and how to effectively integrate hardware and software in the early phase of the design process (*fig. 4*).[7]

6 **T. Nam and W. Lee**, 'Integrating Hardware and Software: Augmented Reality Based Prototyping Method for Digital Products,' *Proceedings of the 2003 Conference on Human Factors in Computing Systems* (New York: ACM Press, 2003).

7 **T. Nam**, 'Sketch-based Rapid Prototyping Platform for Hardware-Software Integrated Interactive Products,' *Proceedings of the 2005 Conference on Human Factors in Computing Systems* (New York: ACM Press, 2005).

facing page

3 **Sketch-based augmented** reality workspace for interactive prototyping

4 **Tangible and augmented** prototype of a mobile device

"Prototyping by Sketching,
Prototyping like Sketching"

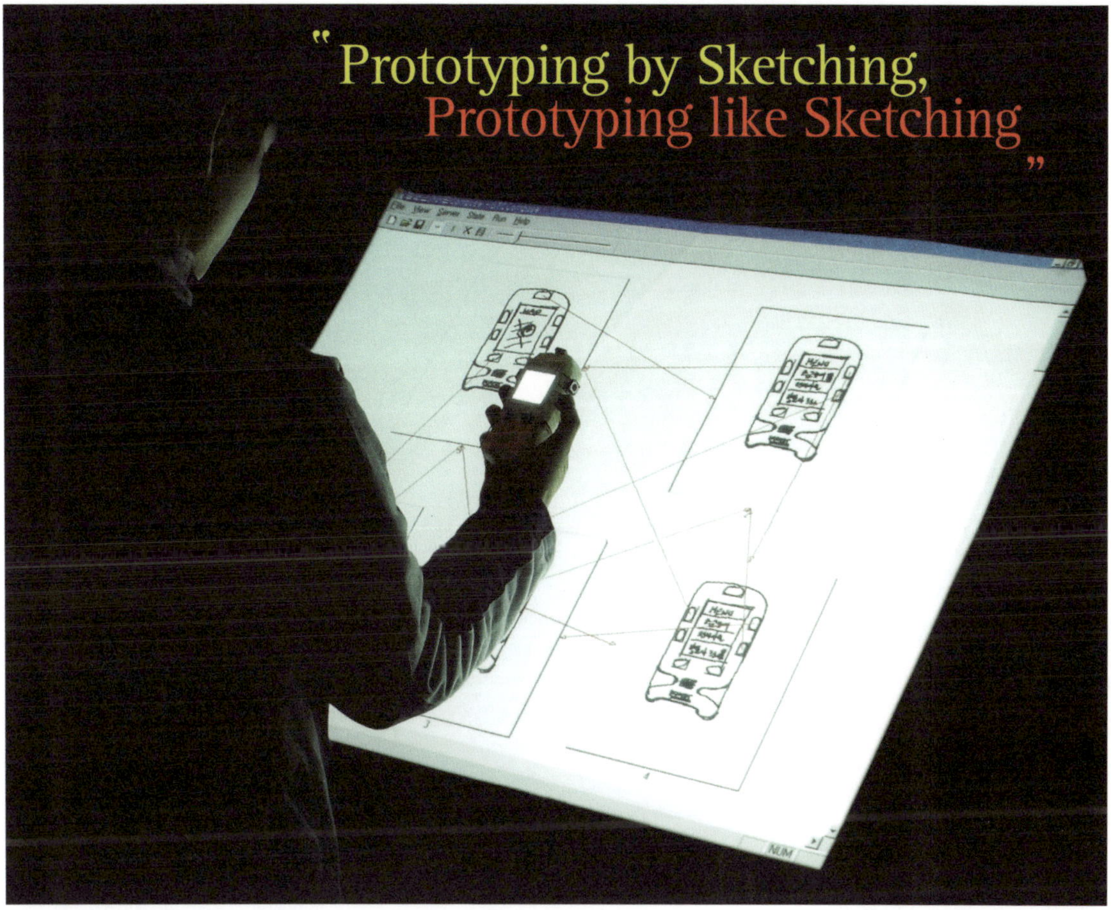

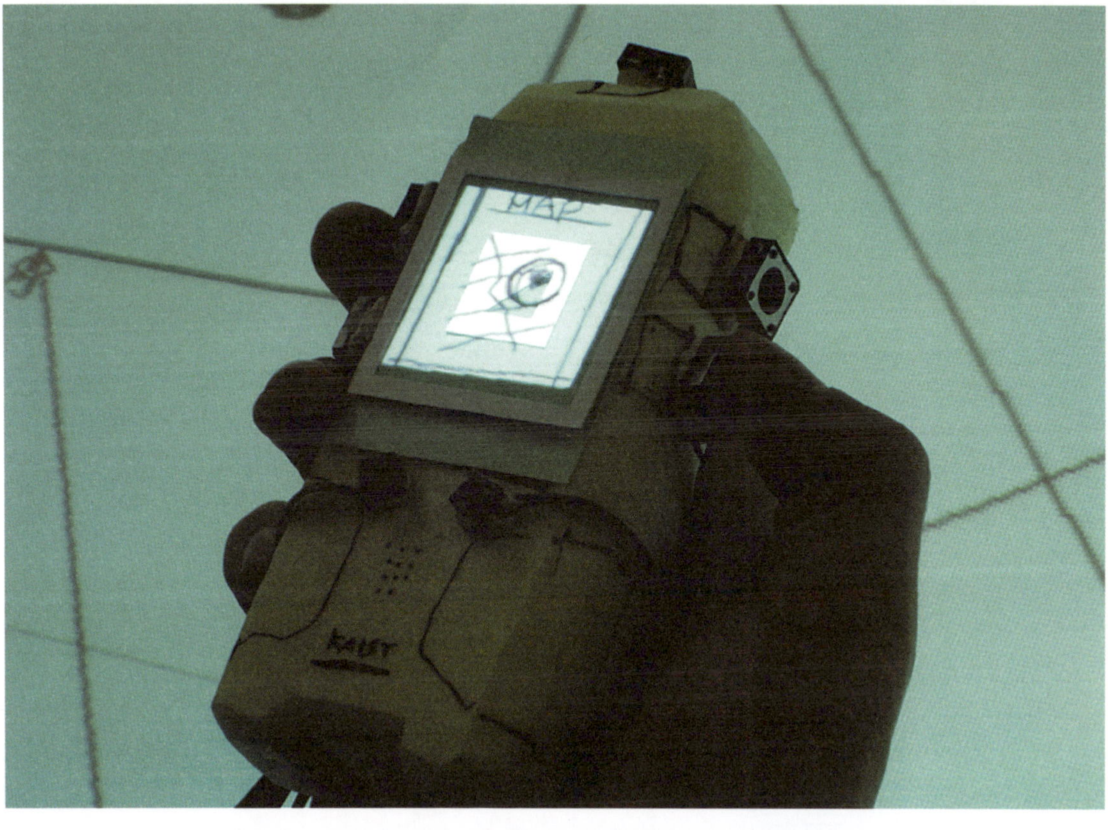

Arduino at Work
the *Hylozoic Soil* control sytem

Robert Gorbet and Philip Beesley
University of Waterloo

Arduino is an open-source physical computing platform that was created to make tools for software-controlled interactivity accessible to non-specialists. The Arduino microcontroller board can read sensors, make simple decisions, and control devices. This palm-sized computing platform is the product of an open-source community project that began with a small group of hardware developers giving workshops and that now numbers many tens of thousands of international users that co-operate in developing specialized applications.

Hylozoic Soil, an interactive environment exhibited in 2007 at the Montreal Museum of Fine Arts, is an example of Arduino at work. The distributed nature of *Hylozoic Soil* and the group behaviour which emerges is strongly related to the open-source Arduino project. Occupants move within the *Hylozoic Soil* structure as they would through a dense thicket within a forest. Microprocessor-controlled sensors embedded within the environment signal the presence of occupants, and motion ripples through the system in response. Dozens of microprocessors, each controlling a series of sensors and actuators, create emergent reactions akin to the composite motion of a crowd. Visitors move freely amidst hundreds of kinetic devices within this environment, tracked by many dozens of sensors organized in 'neighbourhoods' that exchange signals in chains of reflexive responses. The installation is designed as a flexible, accretive kit of interlinking parts organized by basic geometries and connection systems. Variations are created by numerous individuals assembling the work. The result is a turbulent chorus of motion.

The first developers of Arduino—Massimo Banzi, David Cuartielles, Tom Igoe, Gianluca Martino, David Mellis, and Nicholas Zambetti—ran workshops that demonstrated assembly of the devices and gave copies of the board away to stimulate development. A community of developers and users now provides co-operative support, and the programming environment and documentation is written with the neophyte in mind. The Arduino community has to date created myriad documents describing how to extend and interface Arduino with different systems, including

- MaxStream's inexpensive and compact XBee RF wireless transceivers

1 **Two views** of *Hylozoic Soil*, installed at the Montreal Museum of Fine Art, 2007

facing page
2 **Close-up view** of the Printed Circuit Boards used in *Hylozoic Soil*. The Bare-Bones *Arduino* board is mounted to a custom 'daughter' board.

- Bluetooth-enabled mobile phones, with the Arduino BT extended board
- LCD displays
- Cycling 74's Max/MSP/Jitter graphical scripting environment

The following description focuses on the control system that was developed for active functions within the *Hylozoic Soil* project. The micro-controller used in our Arduino platform is an Atmel ATmega168, a tiny computer-on-a-chip that contains specialized hardware to process digital signals, read analog inputs, and communicate over a serial connection. User-designed software is created in a high-level language and programmed into the microcontroller by connecting the Arduino board to a computer's USB port.

The version of the Arduino hardware used for *Hylozoic Soil* is the Bare-Bones Board, Revision C, developed by Paul Badger (www.moderndevice.com). This inexpensive implementation of the platform has a small forty by sixty millimeter footprint, and is provided fully assembled or in kit form. It includes power regulation, timing, and external components for digital inputs and outputs that can control a range of interactive devices. A custom 'daughter board' (or 'shield') was developed to provide three key additional elements to extend the function of the main board: a high-current output stage, configuration switches, and a communication interface. Twelve high-current output channels permit digital control of devices at currents of up to one amp per circuit at voltages up to fifty volts. Twelve switches are read by the software during initialization of the boards and can be used for functions such as configuring individual board addresses and specifying software modes to control individual board behavior. The communication interface converts serial communication signals from the Arduino and supports distribution at high speed to a network of boards using the RS485 standard. The daughter board also provides a sixty-pin ribbon cable interface for connecting actuators and sensing devices, and a two-channel power connector to distribute high currents to actuators as well as a lower current 'electronics' supply.

The *Hylozoic Soil* sculpture includes three kinds of actuator elements: 'breathing' and 'kissing' pore mechanisms actuated by shape-memory alloy 'muscle' wires; 'whisker' elements driven by small direct-current motors; and miniature LED lights. The structural core of *Hylozoic Soil* is a flexible meshwork assembled from small acrylic chevron-shaped tiles that clip together in tetrahedral forms. These units are arrayed into a resilient, self-bracing diagonally organized space-truss. Curving and expanding this trusswork creates a flexible grid-shell topology. Columnar elements extend out from this membrane, reaching upward and downward to create tapering suspension and mounting points. Fitted into this flexible structure are hundreds of small mechanisms that function in ways akin to pores and hair follicles in the skin of an organism.

'Breathing' pores are composed of thin sheets shaped into outward-branching serrated membranes, each containing flexible acrylic tongue stiffeners fitted with monofilament tendons. The tendons pull along the surface

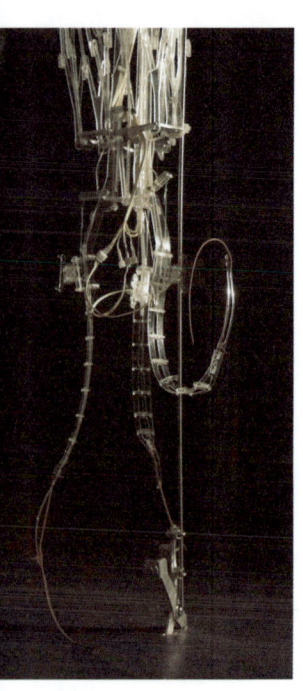

3 **'Kissing' pore** in detail installation view showing actuators driven by muscle wire

facing page
4 **An upward view** of the *Hylozoic Soil* canopy mesh showing a partial network of interconnected microprocessors

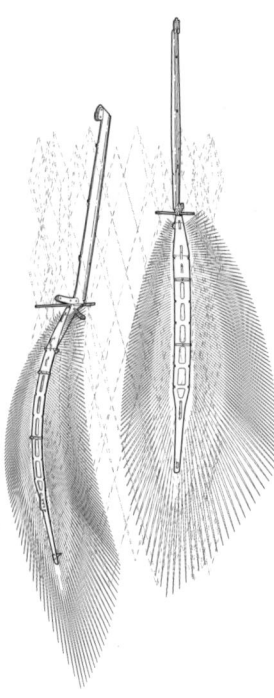

5 **3-D model** of the
   'breathing' pores

1 **The Arduino** can accom-
   modate more, up to eight
   depending on the version,
   but the *Hylozoic Soil* system
   sacrifices some in favour of
   additional digital outputs for
   device control.

of each tongue, producing upward curling motions that sweep through the surrounding air. 'Kissing' pores are a cousin of this mechanism. These use a similar mechanical structure fitted with a fleshy latex membrane and offer cupping, pulling motions. A 'swallowing' pore occurs in a triangular layout that creates a dense series of openings running throughout the meshwork. These openings contain pivoting arms in triangular arrays that push out radially against the surrounding mesh, producing expanding and contracting movements. LED lights are fitted within the lower surfaces of these elements, configured to pulse in synchronization with swallowing motions. 'Whisker' wound-wire pendants are arranged in dense colonies within this environment, supported by acrylic outriggers with rotating bearings and driven by small DC motors. Tensile mounts for the whiskers encourage cascades of rippling, spinning motion that amplify swelling waves of motion within the mesh structure.

Each device is designed to operate at five volts and is interchangeable in the control harness, allowing flexibility in the spatial distribution throughout the meshwork. Under software control, the output drive channels switch current from the high-current five-volt supply to each of the individual actuator elements using a transistor switch. The SMA-actuated pores are driven by ten-inch lengths of 300-micron-diameter Flexinol wire (www.dynalloy.com) that contract when an electrical current runs through them. Mechanical leverage amplifies the half-inch contraction that occurs in each wire and translates this into a curling motion. Whisker elements are composed of flexible wound wire strings extending from the shaft of a small three-pole motor. Yellow LED lights are combined with 150-ohm current-limiting resistors to form a visual actuator configured for the five-volt power supply.

Each daughter board accommodates up to three analog sensors.[1] Sharp infrared proximity sensors with varying detection ranges provide feedback that allows the sculpture to respond to occupant motion. Powered by the five-volt electronics supply, the sensors emit an infrared signal and receive reflected signals from nearby objects, registering the distance of the reflecting surface and feeding that information back to an input on the Arduino board.

The daughter board also contains a communication layer which translates the raw serial data from the Arduino to the RS485 communication standard, and contains jacks to connect the boards to a 'full-duplex, differential multi-drop' bus. RS485 being a differential standard, information is transferred on pairs of wires that carry differing voltages. Bit values are detected by measuring the difference in voltage on the paired wires. This scheme, along with the use of twisted-pair cabling, makes the system less prone to noise-induced communication errors. A full-duplex implementation uses two pairs of wires: one pair for incoming information and the other for outgoing data, allowing for simultaneous communication in both directions along the bus. Each board constitutes one 'drop' of the multi-drop system, and communicates with the others via a single board which assumes the role of 'bus controller'. The Maxim MAX3466 transceiver chip used in the daughter board allows up to 128 such boards to communicate. Since there is the potential for multiple devices to 'drive' the shared bus lines, bus conflicts can occur which result in garbled information at best, and can pose a serious threat to the hardware. The MAX3466 chip includes a pin which allows the microcontroller to effectively 'turn off' the driver circuitry, and this pin is controlled by one of the Arduino's digital outputs.

In addition to the bus transceivers, the daughter board also contains additional hardware which permits simultaneous batch programming of all the devices connected to the bus. Normally, a device is programmed by connecting it to a computer's USB port, then resetting it before running a software tool on the computer to download code to the Arduino. When the Arduino is reset, special code called a 'bootloader' executes for a few seconds, listening for incoming information on the serial port. By setting a switch on the bus controller board to program mode, any board connected to the bus will see messages sent by the computer to the bus controller. If they are all reset just prior to downloading new code from the computer, the bus controller will act as a proxy for all of them in the exchange of information required to download the program, and every board will receive the new code. The bus controller switch is then reset to normal mode and it resumes control of the bus.

The Arduino system combined with the bus architecture described above provides an inexpensive environment for experimentation with distributed intelligence and emergent behaviour in a physical environment. For example, each local board in *Hylozoic Soil* has several layers of response to a presence within the mesh. As a local reflexive response, any board which registers a change in its sensor status immediately activates a reflex device, reinforcing the connection between the actions of the visitor and the sculpture. Reflex responses are followed up by slightly delayed and more orchestrated chains of local reactions, all by devices connected to the triggered board. Additionally, the board informs the rest of the mesh, via the bus controller, that it has detected a visitor. Boards are programmed in software to respond to messages from their spatial neighbours, setting up larger but more muted chains of reaction. A third layer of behavioural control is orchestrated by the bus controller: Since it relays all messages it is aware of the general level of activity within the mesh. It can therefore exercise some control over system-wide behaviour by asking the mesh to set up a general low-level behaviour if things are too quiet, or conversely to quiet down if activity is excessive.

Hylozoic Soil is a project within a body of work that has been gradually moving from individual figures composed of complex hybrid organisms towards immersive architectural environments that behave like highly mobile crowds of interlinked individuals acting in chorus. Recent generations of this work have employed active sensing and actuator mechanisms in pursuit of reflexive, kinetic architectural environments. *Hylozoic Soil* builds upon previous generations by developing a decentralized structure where much of the system is distributed and extensible, based on localized intelligence. The distributed nature of *Hylozoic Soil* and the group behaviour which emerges has much common ground with the Arduino project.

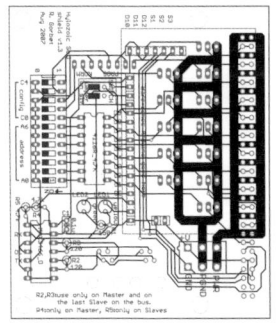

6 **A schematic** of the custom 'daughter' board designed for *Hylozoic Soil*

Technology and Mobile Platforms

Tom Donaldson
Banff New Media Institute

This workshop focused on the use of the Mobile Experience Engine (MEE) in the rapid prototyping and development of rich mobile applications, including locative and proximity-based mobile media applications. Designers and developers wrote application definitions in XML, from which the MEE generated the necessary platform-specific Java or C++ code, compiled and packaged for deployment to mobile devices.

Starting by modifying existing applications, workshop participants quickly became able to create their own locative and proximity-based rich-media mobile applications. As well as becoming familiar with the MEE, a powerful new tool for mobile application development, workshop participants were exposed to a range of new mobile application paradigms to stimulate their own creative exploration, learnt new methodologies for in vivo/in situ design and development, and gained valuable knowledge and confidence to guide future research into mobile applications. More technically expert participants also had the opportunity to explore the architecture of the MEE, understand how to extend its functionality to new devices and new capabilities, and learn how to customize it for specific research or production environments.

facing page
1 **David Gauthier** demonstrating file formats in XML for the Mobile Experience Engine

Biographies
Credits
Index

Biographies

The editors gratefully acknowledge the following individuals who contributed to the Mobile Nation conference and to this anthology.

Matt Adams makes performances, installations, games and interactive artworks. He co-founded Blast Theory in 1991, a group renowned for its multidisciplinary approach pioneering the use of new technologies within performance contexts. The group's work has recently focused on mixed reality, location based games and mobile devices to inspire audiences to question their social relationships. Since 1997, the group has collaborated with the Mixed Reality Laboratory at the University of Nottingham. Works such as *Desert Rain* (1999), *Can You See Me Now?* (2001) and *Uncle Roy All Around You* (2003) have been nominated for four Interactive Arts BAFTA Awards. *Can You See Me Now?* won the Golden NICA for Interactive Art at Prix Ars Electronica 2003.

Julie Andreyev is a Vancouver-based new media artist whose work is influenced by popular entertainment, car cultures, and interactive mobile technologies. Her artistic practice explores the social and spatial character of the city. The most recent projects combine multimedia interactive cars and urban performance. Her work has been shown across Canada, the US, Europe and Japan at venues such as: Viper Festival Basel; SIGGRAPH; ISEA; Media Arts Festival, Tokyo; Elektra Festival, Montreal. Andreyev's work is supported by the Canada Council for the Arts, the British Columbia Arts Council, Foreign Affairs Canada, and the Social Sciences and Humanities Research Council of Canada. She is Associate Professor of Digital Visual Arts at Emily Carr Institute, Vancouver, and co-curator of Interactive Futures Conference, Victoria Canada.
www.fourwheeldrift.com

Philip Beesley practices art and architecture in Waterloo and Toronto, Canada. He is an Associate Professor at the University of Waterloo, School of Architecture in Cambridge, Ontario. He is responsible for the dissemination and publication programs of the Canadian Design Research Network. He co-directs Waterloo's Integrated Centre for Manufacturing, Visualization and Design, a facility combining high-performance computing and automated manufacturing of architectural components. He was educated in architecture at the University of Toronto, in visual art at

Queen's University and in technology at Humber College. Distinctions for his work include the Prix de Rome in Architecture (Canada). Publications include *Fabrication: Examining the Digital Practice of Architecture* (AIA/ACADIA 2004), *Responsive Architectures: Subtle Technologies* (Riverside, 2006), *Future Wood* (Riverside, 2006), and *On Growth And Form: Organic Architecture and Beyond* (TUNS Press 2007).
www.philipbeesley.com

Joanna Berzowska is Assistant Professor of Design and Computation Arts at Concordia University and an active member of the Hexagram Research Institute in Montreal. She is the founder and research director of XS Labs, where her team develops innovative methods and applications in electronic textiles and responsive garments. Her art and design work has been shown in the Cooper-Hewitt Design Museum in New York, the V&A in London, the Millenium Museum in Beijing, various SIGGRAPH art galleries, ISEA, the Art Directors Club in New York, the Australian Museum in Sydney, NTT ICC in Tokyo, and Ars Electronica Center in Linz, among others. She lectures internationally about the field of electronic textiles and related social, cultural, aesthetic, and political issues. She was recently selected for the *Maclean*'s 2006 Honour Roll as one of 'thirty-nine Canadians who make the world a better place to live in.' She received her MSc from MIT for her work titled 'Computational Expressionism' and subsequently worked with the Tangible Media Group of the MIT Media Lab and co-founded International Fashion Machines in Boston. She holds a BA in Pure Math and a BFA in Design Arts.

Jim Budd is Associate Professor in the School of Industrial Design at Carleton University. His research focuses on the methods and technology necessary to support the design and development of interactive products. **Krystina Madej** is a PhD candidate in the School of Interactive Arts & Technology at Simon Fraser University Surrey. Her dissertation focuses on children, narrative and gameplay. **Jenna Stephens-Wells**, **Janice de Jong** and **Laura Mulligan** are fourth-year undergraduate students in the School of Industrial Design at Carleton University with interests in interactive products and technology. **Ehren Katzur** is a first-year undergraduate student in the School of Industrial Design at Carleton University. Katzur has a diploma in Computer

Science from Algonquin College and is interested in the application of computing technology in the field of industrial design.
www.sfu.ca/~krystina/

Barbara Crow has been part of the MDCN since its inception. She has been a core participant in Sampling the Park, EMU (evaluation, mobility and usability), *The Haunting*, and co-editor of *Wi: Journal of the Mobile Digital Commons Network*. As well, she is one of the primary researchers on the Community Wireless Infrastructure Research Project (CWIRP) and former president of the Canadian Women's Studies Association/L'association canadienne des études sur les femmes (CWSA/ACEF).

Steve Daniels is an electronic artist and dumpster diver. He splits his time between Peterborough, Toronto, and the Greyhound. He is a graduate of the Integrated Media program at OCAD and holds an MS in Behavioural Ecology from the University of Manitoba. Steve's practice juxtaposes disparate knowledge systems and experiences in an effort to reveal their underlying structures and assumptions. He is Assistant Professor and Program Director of the New Media option in the School of Image Arts at Ryerson University (Toronto) where he teaches courses in physical computing, telepresence and networked objects.
www.spinningtheweb.org
http://imagearts.ryerson.ca/sdaniels/physcomp/index.html

Marc Davis is Social Media Guru at Yahoo! Inc. where he works on the theory, design, and development of digital media systems that combine contextual metadata and the power of community to enable people to produce, describe, share, and remix media.
http://research/yahoo.com/~marc_davis

Rupinder Deol has been at the Banff New Media Institute as a Mobile Applications Engineer since the beginning of March 2006, where he is passionately working on the Global Heart Rate project. He graduated from the University of Calgary, Canada in 2004 with a Bachelor of Science (Computer Science) degree. He is originally from the wonderful land of India but has been living in this great country, Canada, since 1994. Rupinder loves Canada for its great people and peaceful foreign policies. He loves Banff and its beautiful mountains which he can now visit whenever he likes. Some of his interests are watching movies, hiking

and traveling. He also enjoys being outdoors and visiting art galleries and museums. Drum and violin are his favorite instruments of music and he is hoping to learn how to play them some day.

Sara Diamond is President of the Ontario College of Art & Design, Canada's largest and most diverse art and design university. Diamond is building OCAD's capacities in undergraduate learning, research, and graduate studies, as well as building links with medical and scientific research. Before moving to OCAD in 2005, she was the award-winning Director of Research at the Banff Centre and Artistic Director of Media and Visual Arts for fourteen years. She founded the Banff New Media Institute (BNMI) in 1995 and since then, with her team at Banff and a number of national and international partners, has built the BNMI into a globally recognized content incubator, workshop, and think tank. Diamond's network reaches from Asia to Eastern Europe, Brazil, and the Arctic; from research labs to Silicon Valley; from television to software development. She is currently co-principal investigator on the Mobile Digital Commons Network. Her research and publications explore software visualization and the history of media art.

Since graduating from Cambridge University with a MEng and BA(Hons) in Engineering specializing in electronics and information theory, **Tom Donaldson** has worked as an inventor and entrepreneur, with particular focus on mobile technology and applications and an emphasis on bringing learning from disparate fields into technology innovation. Among other commercial achievements, Tom launched a multi-platform mobile messaging solution later sold to Palm, launched the UK's first mobile entertainment channel, and founded a software company that developed novel artificial intelligence to simplify mobile phone user-interfaces. Among creative achievements, Tom has developed voice-interactive video jewellery shown at the Institute of Contemporary Art in London and the Sydney Opera House and worked on interactive fashion and accessories. In research achievements, Tom has led the engineering in a multi-year, multi-institutional, cross-disciplinary network working on new mobile application development tools for artists and designers and created a wrist-based wireless accelerometer for gestural input to mobile phones. Tom is currently commercializing a new information-free communication technology for the rural areas of the majority world.

Judith Doyle is a filmmaker and new media artist; she is Associate Professor, Faculty of Art, Integrated Media at OCAD. Doyle's 'Warmware' research in collaboration with Baycrest and the Virtual Communities class (team taught with Martha Ladly) was presented recently in Toronto, Oulu, Finland, and Beijing, China.

As CEO of Bight Games, **Stuart Duncan** oversees operations for one of the industry's most innovative mobile game developers. Under his direction Bight's strong R&D focus has propelled them into a leadership position in 3-D mobile game development. Spearheading the bightCode™ game engine and the bightSize™ personalization platform, Duncan has significantly expanded Bight's technological IP while maintaining a strong commitment to the creation and development of world-class mobile games.
www.bightgames.com

Anne Galloway is a social researcher working at the intersections of technology, space and culture. A lecturer and SSHRC Doctoral Fellow in Sociology and Anthropology at Carleton University, Anne is currently finishing her PhD on the social and cultural dimensions of mobility and the design of mobile technologies and locative media for urban public spaces. Anne's research has been presented to a wide variety of international audiences, as well as published in academic journals and industry magazines, and she enjoys teaching undergraduate courses in critical cultural theory and social studies of science and technology. In her spare time Anne can be found hanging out with her cat, reading comics, or writing at www.purselipsquarejaw.org and www.spaceandculture.org

Paula Gardner works in the areas of communication/media studies research, video documentary and mobile design, addressing the relationship between new media technologies and democratic practices of culture. She is Co-Principal Investigator on the project The PORTAGE: Canadian Mobile Experience, which is creating a virtual, interactive street theatre on John Street, Toronto. Gardner is completing a manuscript entitled 'Recovery, Inc.: Depression, Power, Democracy', and a full-length documentary film entitled *Eyes That Don't See, Hearts that Don't Feel*, tracing families fleeing global conflicts during the nineties, and their ongoing displacement due to the American asylum system. See www.mobilelab.ca, for mobile design history experience.

David Gauthier is a programmer from Montreal whose interests lie in advanced imaging and games research. After earning a degree in Mathematics, he worked on several rendering and gaming engines. As part of the Yumi-co collective, he worked on the cuteXdoom game, which has been exhibited in numerous new media festivals. Curious about new ways of interfacing game worlds and experiencing narrative content, David joined the Banff New Media Institute Research Games team and is now focusing on ways to develop avant-garde gaming experiences.

Judy Gladstone has been the executive director of Bravo!FACT (Foundation to Assist Canadian Talent) since 1997. Bravo!FACT (www.bravofact.com) was established in 1995 by CHUM Television's Canadian national cable arts channel Bravo! The foundation is the largest funder of shorts (film and video) in Canada. Thirteen million dollars have been awarded in grants for the production of over a thousand shorts across the country. The shorts are broadcast in Canada on CHUM Television channels, distributed to international broadcasters, and are frequently honoured at film festivals. Bravo!FACT-funded shorts have been screened and have won awards at prestigious festivals around the world, including Cannes, Sundance, and the Toronto International Film Festival. Prior to her arrival at CHUM, Gladstone was coordinator of the CIDA-funded Canada Fund for Dialogue and Development in the Middle East. Ms. Gladstone's education includes a graduate degree from the Sorbonne, France, and an undergraduate degree from Université Laval (Québec City).

Robert Gorbet is Assistant Professor of Electrical & Computer Engineering at the University of Waterloo, with cross-appointments to Mechanical Engineering and the School of Architecture. He holds BASc (1992), MASc (1994) and PhD (1997) degrees from the University of Waterloo. He is also a practicing technology artist, and has exhibited technology-mediated works internationally since 2002 in collaboration with artists, designers and architects. He is an award-winning instructor, teaching courses in professionalism and ethics, microcontroller interfacing, and robotics. In 2004 he helped develop Technology Art Studio, a course combining engineering and sculpture students in interdisciplinary project groups to create technology-mediated sculptural works.
http://ece.uwaterloo.ca/People/faculty/gorbet.html
www.gorbetdesign.com

Since his media career began in India in 1983, **Nathon Gunn** has built an international reputation for pioneering innovation in the fields of media, entertainment and technology. Among many firsts, Gunn helped launch the interactive divisions at Miramax, Chum, and Universal Studios. International media firm Bitcasters, which Nathon co-founded in 1996, has produced award-winning film, television, Internet, and computer game projects for clients including Disney, the Family Channel, MuchMusic, and more. A published author and prolific speaker, Gunn has also advised organizations ranging from the Charles Bronfman Foundation to the office of former prime minister Paul Martin.

Drew Hemment has recently taken up a post at Imagination@Lancaster, a new interdisciplinary research institute at Lancaster University. Artistic Director of the Futuresonic festival, which he founded in 1995. Director of Future Everything, a non-profit creative 'Community Interest Company.' Member of the collaborative Loca group, which he developed during an AHRC Research Fellow in Creative Technologies at Salford University. Founder member of PLAN, the Pervasive and Locative Arts Network, funded by EPSRC. Curator of numerous exhibitions on media art, mobile culture, and locative media. DJ and event organizer at reggae blues warehouse parties and clubs in the early UK dance scene in the late eighties. Completed a PhD at University of Lancaster, and an MA (Distinction) at the University of Warwick, when he was a participant in the Virtual Futures events.

Bruce Hinds is Assistant Professor of Design at the Ontario College of Art & Design where he teaches Design Process, Interaction Design, Design Drawing, Think Tank (co-chair) and Biomimetics (curriculum leader). As a licensed architect, Bruce maintains an active practice addressing issues of sustainable community structures in the Third World. Current projects include working with a multidisciplinary team of physicians and specialists in the Kilimanjaro Region of Tanzania to construct a sustainable community for children affected and infected with HIV. Bruce is an active member of the Architectural Institute of British Columbia, the Ontario Association of Architects, the Royal Architectural Institute of Canada, He is also associate of the Architectural Institute of America, member of Architects for Humanity and the Toronto Society of Architects and an associate of the Ontario College of Art. Bruce holds degrees in psychology, architecture and painting.

Daniel Jolliffe is a media and visual artist whose work blends sculptural, conceptual and technological approaches. His recent project One Free Minute, a mobile sculpture designed to allow for instances of anonymous public speech, received publicity and online coverage in over a dozen languages. Most recently his work was included in the 2006 International Symposium of Electronic Art, the Civil Rights Censorship entry of the 'Yahoo! Directory' and Design Life Now, the triennial exhibition of American design at the Cooper-Hewitt Museum. He is Canadian and lives in Vancouver, where he currently teaches in the School of Interactive Arts and Technology at Simon Fraser University Surrey.
www.arduino.cc
http://downloads.oreilly.com/make/arduinoMAKE07.pdf

James E. Katz PhD, is Chair of the Department of Communication at Rutgers University where he also directs the Center for Mobile Communication Studies. Currently he is investigating how personal communication technologies, such as mobile phones and the Internet, affect social relationships and how cultural values influence usage patterns of these technologies. His main current research involves the Liberty Science Center museum project to look at teen use of mobile communication for informal science learning (sponsored by the US National Science Foundation). Katz has had a distinguished career researching the relationship among the domains of science and technology, knowledge and information, and social processes and public policy. His books include *Perpetual Contact: Mobile Communication, Private Talk and Public Performance* (co-edited with Mark Aakhus), *Connections: Social and Cultural Studies of the Telephone in American Life*, and *Social Consequences of Internet Use: Access, Involvement, Expression* (co-authored with Ronald E. Rice). He is the author of more than forty peer-reviewed journal articles and his works have been translated into five languages and re-published in numerous edited collections.

Associate Professor **Filiz Klassen** of Ryerson University is the co-editor of *Transportable Environments 3*, the third book on portable architecture and design published by Spon Press, UK. Klassen is the recipient of a research/creation grant from the Social Sciences and Humanities Research Council of Canada for her project entitled 'Malleable Matter.' Scheduled to be exhibited in 2008, this project involves a life-size architectural installation of building components such as walls, ceilings, and furniture that makes creative use of textiles and related materials innovations.

Martha Ladly is Mobile Nation Conference Leader, and Associate Professor of Design at Ontario College of Art & Design, specializing in interactive communication. Ladly is a senior researcher with the Mobile Digital Commons Network, and engages in teaching and mentorship outside of the OCAD community with the Canadian Film Centre's Interactive Project Lab and Interactive Art and Entertainment Programs. In previous lives, Ladly directed Horizonzero.ca at the Banff New Media Institute, worked with Peter Gabriel as Head of Design for his Real World Group in the UK, and played keyboards with Toronto new wave band Martha and the Muffins. She is currently pursuing graduate studies in the joint Communication and Culture program at York University.

Angus Leech is a new media artist, writer, and editor. He is currently Lead Artist/Producer for the Banff New Media Institute's ART Mobile Lab, a research unit that develops new multimedia content and software applications for handheld mobile media devices and conducts user-centered research to investigate patterns of mobile media use in outdoor environments. As part of the Mobile Digital Commons Network, the lab's current research initiative is *Tracklines*, a mediascape project which explores the potential of locative media for enriching the experience of trail environments in Banff National Park.

Maroussia Lévesque holds an undergraduate degree in computation arts from Concordia University. Her curriculum is oriented towards new technologies and political science, as she is interested in the politics of computers. She has worked in grassroots organizations in Canada and Brazil and is motivated by the potential of subcultures as social emancipators. She is the recipient of the Mary Higgins Bursary, and won the first prize of Jeunes Critiques en Arts Visuels. Her work has been published in *Le Devoir*. She joined Obx Labs in 2005 and is currently its conceptual lead. Lévesque is learning her fifth language and counting.
www.digital-spa.com
www.elaborate.ca

Jason Lewis is a poet, digital media artist, and software designer. His research/creation practice revolves around experiments in visual language, text, and typography, with a core interest in how the deep structure of digital media can be used to create innovative forms of expression. His creative work and his writing about new media have been presented at conferences, festivals, and exhibitions internationally.

His work has been supported by Canadian Heritage, the Social Sciences and Humanities Research Council, Fonds québécois de la recherche sur la société et la culture, Hexagram, Arts Alliance, Canada Council for the Arts, and Arts Council England. He is currently an assistant professor of computation arts at Concordia University, where he founded Obx Laboratory for Experimental Media.
www.obxlabs.net

Michael Longford is Chair and Associate Professor in the Department of Design and Computation Arts at Concordia University in Montreal. His creative work and research activities reside at the intersection of photography, graphic design, and new media. Currently, he is co-principal investigator for the Mobile Digital Commons Network (MDCN), a joint research project launched by Concordia University, the Ontario College of Art & Design, and the Banff New Media Institute. He has organized numerous workshops, artist talks, exhibitions, and conferences. He is a founding member of Hexagram Institute for Research and Creation in Media Arts and Technologies and belongs to the Advanced Digital Imaging and 3-D Rapid Prototyping Group.

David McIntosh is Associate Professor, Media Studies, at the Ontario College of Art & Design. He holds a PhD in Communications and Culture from York University. His research interests include globalization and the political-economies of audiovisual spaces; network theories and practices; new media narrativity; rapid prototyping; Latin American media studies; queer media; and insurrectional media history. He is faculty researcher with the Mobile Digital Commons Network and, with Michael Longford, is Research and Creative Director of the mobile media game *The Haunting*. His critical writing on new media has been published extensively in books and periodicals, and he has curated programs of new media for the National Gallery of Cuba, the National Gallery of Argentina and Cinemathèque Ontario.

Shawn Micallef lives in Toronto and is co-founder of the location-based mobile phone documentary project [murmur], Associate Editor and feature writer at *Spacing Magazine*, and co-founder of the Toronto Psychogeography Society, a group of *flâneurs* who drift through and explore the city. Shawn has an MA in Political Science and was a resident at the Canadian Film Centre Media Lab where [murmur] was initially developed. Shawn's writing has been found in the *Globe and Mail*, the *National Post*, and

Maisonneuve Magazine as well as *Eye Weekly*, where his 'Stroll' column led him to all corners of Toronto. Shawn is also an instructor at the Ontario College of Art & Design.
www.spacing.ca | www.murmur.info
www.psychogeography.ca

Tek-Jin Nam is Associate Professor in the Department of Industrial Design at KAIST in Korea. He received his BS and MS in Industrial Design from KAIST and received his PhD from Brunel University, UK. His primary interests lie in human centered innovation for interactive products and systems, collaborative design, tangible interaction design and interactive prototyping. He is a member of the Scientific Committee of the *International Journal on Interactive Design and Manufacturing* and an Executive Director of Korea Society of Design Science. He has led design and research projects for Samsung Electronics, LG Electronics, the Korean Ministry of Education and Human Research Development, SK Telecom, KTF, and others.
http://cidr.kaist.ac.kr/midas | http://cidr.kaist.ac.kr

Leena Saarinen, MA, is a practitioner and researcher of interactive storytelling. She has participated in several collaborative media art productions including award-winning concepts Villa Mirdia chat world and *Accidental Lovers*, a participatory black comedy for television. As a founding member of virtual theatre group Avatar Body Collision she is exploring new forms of participatory drama and live online performance. Currently Saarinen is working on her Doctor of Arts dissertation 'Scripting for Computational Drama,' which collides storytelling genre rules with software metadata and ontology to discover useful ways to write and produce interactive stories.

Kim Sawchuk, PhD, is Co-Director of the Mobile Media Lab at Concordia University. As a member of the Mobile Digital Commons Network, she has been a core participant in EMU (evaluation, mobility and usability), *The Haunting*, and is a co-editor of *Wi: Journal of the Mobile Digital Commons Network* (http://wi-not.ca). Sawchuk is Associate Professor in the Department of Communication Studies at Concordia University and the current editor of the *Canadian Journal of Communication* (www.cjc-online.ca).

Gabe Sawhney is a hacker working at the edges of code and culture. He is co-creator of [murmur], a location-specific oral storytelling project that makes accessible the hidden stories of cities in Canada, the US and the UK. He is co-founder of Wireless Toronto, a community group offering free-to-use hotspots in public and publicly-accessible spaces in the city, each featuring its own 'hyperlocal' community portal. Gabe is involved with several other web, locative, video and installation projects bridging art, politics and technology. His heart rests firmly with the simple, the intuitive, and the cheap.

Thecla Schiphorst is a media artist and Associate Professor in the School of Interactive Arts & Technology at Simon Fraser University in Vancouver, Canada. She is Director of the whisper[s] research group, and her media art installations have been exhibited internationally in Europe, Canada, the US and Asia in many venues including Ars Electronica, the Dutch Electronic Arts Festival (DEAF), Future Physical, SIGGRAPH, Interaction '97, the Wexner Centre for the Arts, the Montreal New Media Festival, ISEA, the School of the Visual Arts in New York, the Canadian Cultural Centre in Paris, and Dance Umbrella at the London ICA.

Parmesh Shahani is heading a new ideas incubation lab in Bombay, India, for Mahindra and Mahindra, one of India's largest business conglomerates, with a significant presence in key sectors of the Indian economy. He also serves as India-based Research Affiliate for the Convergence Culture Consortium (C3) at MIT, a consortium he helped co-found in 2005. Shahani holds a Master's in Comparative Media Studies from MIT and two Bachelors degrees from the University of Mumbai. His previous work experiences include founding and managing India's first Internet youth portal, Freshlimesoda.com, business development for Sony Entertainment Television, writing for *Elle* magazine and the *Times of India* newspaper, helping make a low-budget feature film and teaching as a visiting faculty member at a Bombay college. Mr. Shahani is the author of the forthcoming book *Disco Jalebi: An Ethnography of Gay Bombay* and his academic interests include globalization, media convergence, identity formation, online communities, fashion, youth culture, lifestyle, branding, South Asian pop culture and Bollywood cinema.
parmesh@mit.edu.

Leslie Sharpe is AT&T Fellow, Assistant Professor, and Area Head of Digital Art at Indiana University Bloomington. She employs ghost genres in projects using mobile devices, Bluetooth, podcasting and locative media to explore subjectivity, embodiment, social networks and place. Sharpe has exhibited in Canada, the US, and Europe including Kiasma, Artists Space, Exit Art, and PS1, where she was an artist in residence, and has written for Leonardo's issue on locative media. Sharpe has lived in Toronto, New York, and San Diego, and received her MFA at UCSD where she was also Faculty Fellow. Sharpe was born in Medicine Hat, Alberta and lives in Alberta, when not teaching. http://lesliesharpe.net

Geoffrey Shea is Assistant Professor at the Ontario College of Art & Design, where he has taught new and emerging media for artists and designers since 1986. Currently he is co-leading a research team developing a broad locative environment, a streetscape virtual theatre triggered by mobile users. Shea's artwork has been exhibited and collected widely and he recently released a DVD/CD of his musical work with Gigi Minor. www.unscrambled.com

By focusing on shopping malls, markets, theme parks, tourist attractions, and high-tech facilities, **Rob Shields**'s research seeks insights into how understandings of social space, the metropolis, and online culture impact identity and sociability, pleasure and taste, the cultures of institutions and cities, and 'knowledge' and 'innovation.' This project has been extended through an interdisciplinary journal *Space and Culture* (Sage) and publications on the spatiality of *Places on the Margin* and on consumption as *Lifestyle Shopping*. Recent research considers *The Virtual*, how construction innovations are literally *Building Tomorrow* (André Manseau co-editor, 2005) and what can be learned about cultural capitals and about *The Urban after Katrina* from New Orleans and other recovering cities. www.spaceandculture.org

Suzanne Stein leads the Mobile Games Group for SMARTlab, and co-supervises the PhD cohort working within Peoplelab. She has been a member of the Canadian Film Centre Media Lab since 1997 and was a co-mentor for the Interactive Project Lab, guiding and nurturing innovative technology projects for market launch. She was Discipline Lead of Research Division, Experience Modeling, at Sapient, as well as co-lead for its User Experience Group. More recently, she led Nokia's foresighting capabilities for strategic and creative ends. She has served as a juror and moderator for the WSA, a United Nations task group concerned with the information society.

Maria Stukoff is an artist and curator working in the field of interactive media art. Stukoff contributes widely to the digital entertainment and emergent technologies debate, most notably as Programme Director of the Game Alliance, an independent network for game

studios in the northwest of England, and by lecturing in the Narrative and the Moving Image programme at the International Centre for Digital Content, Liverpool John Moore's University. Currently attending to PhD research at the Manchester Metropolitan University entitled 'Mobile and Wireless Networks as Public Art'. Her recent commissions explore proximity-based Bluetooth environments using mobile telephony.

Minna Tarkka is director of m-cult, centre for new media culture in Helsinki. She has been involved in setting up several organisations, educational programs and events of media art and culture, including the MA in New Media at the University of Art and Design, ISEA'94 and ISEA2004. At m-cult, her work has been to advocate, document and communicate practices of media culture, with a special focus on participatory, urban and mobile media. Her research and writing aims at a critical study of new media, creativity, and participation. www.m-cult.org

Nigel Thrift is Vice-Chancellor of the University of Warwick. Prior to this he was Pro-Vice-Chancellor for Research at the University of Oxford. Thrift was made Head of the Division of Life and Environmental Sciences at Oxford in 2003, before which he chaired the Research Committee at The University of Bristol (2001–2003) and Bristol's Research Assessment Panel (1997–2001). Thrift was born in Bath, educated at Aberystwyth and Bristol, and is an international research figure in the field of geography (one of the top five most cited geographers in the world in the SSCI/ACHI Indexes, 1988–2002). During his academic career Professor Thrift has been the recipient of a number of distinguished academic awards including the Royal Geographical Society Victoria Medal for contributions to geographic research in 2003. Nigel Thrift is an Academician of the Academy of Learned Societies for the Social Sciences, and was made a Fellow of the British Academy in 2003. He currently chairs Main Panel H of RAE 2008 from 2003–2006; was a member of the Panel for Geography for the RAE 2001; has been a member of the Leverhulme Prize Fellowship Geography Panel since 2000 and was a member of the ESRC Research Priorities Board between 2001 and 2005. He is Visiting Professor of Geography at the University of Oxford and an Emeritus Professor of Geography at the University of Bristol.

Vincent John Vincent is the co-founder (with Francis MacDougall) and creative force behind GestureTek Inc. (formerly Vivid Group), which is the inventor of and world leader in camera-enabled gesture control of computer technology and displays over the past twenty-plus years. From the multi-patented Video Gesture Control (VGC) technology which spawned the GestureXtreme system and software applications in entertainment, Vincent, as GestureTek's president, has been instrumental in moving the company forward into new technological and marketing directions: GestPoint ('touchless' point-control of screen), GestureFX (interactive floor, wall, and table displays) and IREX (rehabilitative technology). Vincent has overseen over 2,000 public installations worldwide, and there are multiple consumer licenses for PC, console, and toy markets. GestureTek Mobile is their latest development.

www.gesturetek.com

David Vogt is a technology innovator and entrepreneur with solid corporate, academic, and R&D experience. Vogt is Director of Digital Learning Projects at UBC and champions a set of advanced collaborative R&D projects in new media and mobile media technologies, including the Mobile MUSE Network and GUSSE. Vogt also shepherds other early stage ventures and is an active contributor to a number of private and public boards. His family life centres on awesome wife Tracy and four amazing children.

Nina Wakeford is Director of INCITE and Reader in Sociology at Goldsmiths, University of London. She founded INCITE in 2001 at the University of Surrey after noticing the increasing use of qualitative research methods, in particular ethnography, in the new technology design field. Researchers at INCITE have been funded by Intel, Sapient, NSF, FujiXerox Palo Alto and the UK government. In 2005 Nina led a government 'mission' on the future of user-centred design. Currently she holds a three-year Economic and Social Research Council Fellowship which focuses on the use of social research in art and design. As part of this fellowship she is exploring the potentials of developing collaborative and studio-based practices for sociology, as well as the use of exhibitions and installations for research translation.

www.studioincite.com | www.goldsmiths.ac.uk

Ron Wakkary is Associate Professor in the School of Interactive Arts & Technology at Simon Fraser University in British Columbia. His research interests lie in the design of ubiquitous computing systems including responsive environments, personal technologies, and tangible user interfaces, and the study of interaction design related methods and practice.

He is currently researching the concept of everyday design, funded by the Social Sciences and Humanities Research Council, and co-leads the Interactivity theme in the Canadian Design Research Network. He led the Am-I-able Network for Responsive and Mobile Environments, a Canadian research network in mobile, wearable, and responsive technologies.

Eric Zimmerman has been working in the game industry for more than twelve years. He is the co-founder and CEO of Gamelab (www.gamelab.com), a game development company based in New York City that focuses on experimental and innovative games. Gamelab creates and self-publishes award-winning single player and multiplayer games that are distributed online, on mobile phones, and through retail, including the hit downloadable game *Diner Dash*. Pre-Gamelab titles include *SiSSYFiGHT 2000* (www.sissyfight.com) and the PC title *Gearheads*. Zimmerman has taught courses at Massachusetts Institute of Technology, New York University, and Parsons School of Design. He has lectured and published extensively about game design and game culture and is the co-author with Katie Salen of *Rules of Play: Game Design Fundamentals* (MIT Press, 2004) and *The Game Design Reader: A Rules of Play Anthology* (MIT Press, 2006), as well as the co-editor with Amy Scholder of *RE:PLAY: Game Design and Game Culture* (Peter Lang Press, 2004).

www.ericzimmerman.com

Jan-Christoph Zoels is responsible for user experience design at Experientia, an experience design consultancy based in Turin, Italy. Until recently he was Senior Associate Professor at Interaction Design Institute Ivrea, where he ran the business innovation workshops called Applied Dreams. In his work Zoels focuses specifically on people's experience of mobile services and applications, and on using information technology to support simplicity. Previously he was Director of Information Architecture for Sapient (New York), and senior designer at Sony Design Center USA, responsible for strategic product development. He holds four patents.

www.experientia.com

Image Credits

Credits

MOBILE NATION BOOK

EDITORS
Martha Ladly
Philip Beesley

PUBLICATION COORDINATOR
Siobhan O'Flynn

COPY EDITOR
Leah Sandals

PRODUCTION DIRECTOR
Philip Beesley

ART DIRECTION AND COORDINATION
Eric Bury

DESIGN AND ART DIRECTION
Nevena Niagolova
Fiona Chung
Mary Christine Plaza

MDCN RESEARCHERS

MDCN RESEARCHERS
BANFF NEW MEDIA INSTITUTE
Davide Di Saro
Sarah Hoyt
Susan Kennard
Angus Leech
Christopher Quine

CONCORDIA UNIVERSITY
Jason Lewis
Kim Sawchuk

ÎLE SANS FIL [ISF]
Philippe April
Benoit Grégoire
Daniel Lemay
Michael Lenczner

ONTARIO COLLEGE OF ART AND DESIGN
Paula Gardner
Bruce Hinds
Martha Ladly
David McIntosh
Geoffery Shea

YORK UNIVERSITY
Barbara Crow

MDCN RESEARCH ASSISTANTS
André Arnold
Yannick Assogba
Neil Barratt
Lucie Belanger
Lysanne Bellemare
Jeff Bolingbroke
David Bouchard
Hugues Bruyere
Thibaut Duverneix
Anna Friz
Jennifer Gabrys
Alison Harvey
Geoffrey Jones
Wai Kok
Marie-Claude Landry
Ganaele Langlois
Janice Leung
Ken Leung
Maroussia Levesque
Zehuan Liu
Raed Moussa
Anton Nazarko
Nevena Niagolova
Leanna Palmer
Marit-Saskia Wahrendorf
Elie Zananiri

MDCN LEAD ENGINEER
Tom Donaldson

MDCN ENGINEERS
Amitava Biswis
Rupinder Deol
Armen Forget
David Gauthier
Sukhmeet Singh
Jagmit Singh
Alexander Taler

MDCN COORDINATORS
Patricio Davila
Brenda Goldstein
Cindy Schatkoski
Andrea Zeffiro

MDCN Student Interns

Adam Brandejs
Fiona Chung
Amanda Cooley
Nigel Craig
Jérôme Delapierre
Jan Drewniak
Garry Ing
Benjamin Lemar
Connie Leung
John Pavagic
Mary Christine Plaza
Mark Poon
Bryn Reed-Ludlow
Philip Sportel
Peter Todd
Jennie Ziemianin

MDCN Partners

Bravo!FACT
Hexagram Institute for Research
Creation in Media Arts & Technologies
Networks for Emerging
Wireless Technologies (NEWT)

MOBILE NATION CONFERENCE CREDITS

Partners

Ontario College of Art & Design
Mobile Digital Commons Network
Canadian Design Research Network

Sponsors

Canadian Heritage
Social Sciences and Humanities
Research Council
Networks of Centres of Excellence
Riverside Architectural Press
Waterloo Architecture Cambridge
Interactive Project Lab
Polycom
Backbone Magazine

CONFERENCE ADMINISTRATION

Principal Investigators

Sara Diamond
Michael Longford

Conference Leader

Martha Ladly

Conference Coordinator

Anthea Foyer

Assistant Coordinator

Michelle Wolfenden

CDRN INTERACTIVE AND SENSING TECHNOLOGIES WORKSHOPS

Coordinator

Ron Wakkary

Workshop Administration

Kevin Muise
Greg Corness

Workshop Leaders

Tom Donaldson
Daniel Jolliffe
Tek-Jin Nam

Index